# Environmental Citizenship in the Indian Ocean Region

The scale and severity of our environmental challenges are quickly becoming apparent. The Indian Ocean region features many places particularly vulnerable to the effects of environmental degradation and climate change, which will have profound social, economic, and cultural impacts. The increasing preoccupation with the state of the environment is also having significant political effects, including on the concept and content of citizenship. The language of citizenship has permeated environmental discourse and, conversely, environmental issues are often articulated in the language of citizenship.

This book explores environmental citizenship and civil society responses to environmental challenges in the Indian Ocean region. The articles provide practical insights to improve resilience and adaptation, as well as conceptual insights into the nature of environmental citizenship discourse and practice across this vast region, from Mauritius to Malaysia. The volume showcases the complex field of environmental citizenship through a wide range of approaches, and alongside closely related concepts, such as environmental governance, environmental education, environmental justice, and corporate social responsibility. In essence, the book provides a rich, diverse, and multidimensional picture of environmental citizenship in the Indian Ocean region.

The chapters in this book were originally published as a special issue in the *Journal of the Indian Ocean Region.*

**Benito Cao** is Senior Lecturer in Politics at the University of Adelaide, Australia. He has a vast teaching experience and has received numerous teaching awards. He has published in the fields of identity politics, critical thinking, citizenship studies and environmental politics, and is the author of *Environment and Citizenship* (Routledge 2015).

## Transforming Environmental Politics and Policy

*Series Editors*
*Timothy Doyle, Keele University, UK and University of Adelaide, Australia and Philip Catney, Keele University, UK*

The theory and practice of environmental politics and policy are rapidly emerging as key areas of intense concern in the first, third and industrializing worlds. People of diverse nationalities, religions and cultures wrestle daily with environment and development issues central to human and non-human survival on the planet Earth. Air, Water, Earth, Fire. These central elements mix together in so many ways, spinning off new constellations of issues, ideas and actions, gathering under a multitude of banners: energy security, food sovereignty, climate change, genetic modification, environmental justice and sustainability, population growth, water quality and access, air pollution, mal-distribution and over-consumption of scarce resources, the rights of the non-human, the welfare of future citizens-the list goes on.

What is much needed in green debates is for theoretical discussions to be rooted in policy outcomes and service delivery. So, while still engaging in the theoretical realm, this series also seeks to provide a 'real world' policy-making dimension. Politics and policy making is interpreted widely here to include the territories, discourses, instruments and domains of political parties, non-governmental organizations, protest movements, corporations, international regimes, and transnational networks.

From the local to the global-and back again-this series explores environmental politics and policy within countries and cultures, researching the ways in which green issues cross North-South and East-West divides. The 'Transforming Environmental Politics and Policy' series exposes the exciting ways in which environmental politics and policy can transform political relationships, in all their forms.

**Contesting Global Environmental Knowledge, Norms, and Governance**
*Edited by M. J. Peterson*

**The Politics of Arctic Resources**
Change and Continuity in the "Old North" of Northern Europe
*Edited by E. Carina H. Keskitalo*

**Environmental Blockades**
Obstructive Direct Action and the History of the Environmental Movement
*Iain McIntyre*

**Environmental Citizenship in the Indian Ocean Region**
*Edited by Benito Cao*

For more information about this series, please visit: https://www.routledge.com/Transforming-Environmental-Politics-and-Policy/book-series/TEPP

# Environmental Citizenship in the Indian Ocean Region

*Edited by*
**Benito Cao**

Routledge
Taylor & Francis Group
LONDON AND NEW YORK

First published 2021
by Routledge
2 Park Square, Milton Park, Abingdon, Oxon, OX14 4RN

and by Routledge
605 Third Avenue, New York, NY 10158

*Routledge is an imprint of the Taylor & Francis Group, an informa business*



*British Library Cataloguing-in-Publication Data*
A catalogue record for this book is available from the British Library

ISBN13: 978-0-367-75777-9 (hbk)
ISBN13: 978-0-367-75779-3 (pbk)
ISBN13: 978-1-003-16396-1 (ebk)

Typeset in Myriad Pro
by codeMantra

**Publisher's Note**
The publisher accepts responsibility for any inconsistencies that may have arisen during the conversion of this book from journal articles to book chapters, namely the inclusion of journal terminology.

**Disclaimer**
Every effort has been made to contact copyright holders for their permission to reprint material in this book. The publishers would be grateful to hear from any copyright holder who is not here acknowledged and will undertake to rectify any errors or omissions in future editions of this book.

# Contents

# Citation Information

The chapters in this book were originally published in the *Journal of the Indian Ocean Region*, volume 16, issue 1 (March 2020). When citing this material, please use the original page numbering for each article, as follows:

**Introduction**

*Environmental citizenship in the Indian Ocean region*
Benito Cao
*Journal of the Indian Ocean Region*, volume 16, issue 1 (March 2020) pp. 1–5

**Chapter 1**

*Climate change adaptation: the need for an Indian Ocean regional metamorphosis*
Victor R. Savage and Lin Qi Feng
*Journal of the Indian Ocean Region*, volume 16, issue 1 (March 2020) pp. 6–26

**Chapter 2**

*Indian Ocean fisheries regulation: exploring participatory approaches to support small-scale fisheries in six States*
Erika Techera
*Journal of the Indian Ocean Region*, volume 16, issue 1 (March 2020) pp. 27–46

**Chapter 3**

*Ecosocial work through corporate social responsibility: the case of company engagement with civil society in coastal communities in Mauritius*
Brita Backlund Rambaree
*Journal of the Indian Ocean Region*, volume 16, issue 1 (March 2020) pp. 47–62

**Chapter 4**

*Crisis, constitutionalism and the geographies of belonging: Indian eco-politics in the Anthropocene*
Ramya K. Tella
*Journal of the Indian Ocean Region*, volume 16, issue 1 (March 2020) pp. 63–78

**Chapter 5**

*Contradictions of citizenship and environmental politics in the Arabian littoral*
Ian R. Simpson
*Journal of the Indian Ocean Region*, volume 16, issue 1 (March 2020) pp. 79–99

**Chapter 6**

*Environmental citizenship for inclusive sustainable development: the case of Kelab Alami in Mukim Tanjung Kupang, Johor, Malaysia*
Serina Rahman
*Journal of the Indian Ocean Region*, volume 16, issue 1 (March 2020) pp. 100–118

# Contributors

**Benito Cao** Department of Politics and International Relations, The University of Adelaide, Australia.

**Lin Qi Feng** School of Humanities, Nanyang Technological University, Singapore.

**Serina Rahman** ISEAS-Yusof Ishak Institute, Singapore.

**Brita Backlund Rambaree** Department of Social Work and Criminology, University of Gävle, Sweden.

**Victor R. Savage** S. Rajaratnam School of International Studies, Nanyang Technological University, Singapore.

**Ian R. Simpson** Department of Heritage and Society, Leiden University, the Netherlands.

**Erika Techera** UWA Law School and UWA Oceans Institute, The University of Western Australia, Perth, Australia.

**Ramya K. Tella** Department of Geography, King's College London, UK.

INTRODUCTION

# Environmental citizenship in the Indian Ocean region

Benito Cao

The scale and severity of the environmental challenges we face are quickly becoming apparent. The Indian Ocean region features many places that are particularly vulnerable to the effects of environmental degradation and climate change, which will have profound social, economic, and cultural impacts. The increasing preoccupation with the state of the environment is also having significant political effects, including on the concept and content of citizenship. The language of citizenship has permeated environmental discourse and, conversely, environmental issues are often articulated in the language of citizenship. Thus, for example, states around the world are recognizing environmental rights; activists and governments refer to our duties to the environment; and corporations present themselves as good environmental citizens; all the while citizens are exhorted to 'reduce, reuse, and recycle'. These three political actors – states, citizens, and corporations – are being drawn into addressing our serious environmental predicament. In the process, traditional formulations of citizenship are being deployed and transformed, and new ones are gradually emerging (Cao, 2015, pp. 72–103); and scholars have begun to consider the political implications of environmental challenges and ecological values under a new field of study called environmental citizenship.

This special edition of the *Journal of the Indian Ocean Region* (JIOR) focuses on environmental citizenship and civil society responses to environmental challenges in the Indian Ocean region. The significance of environmental citizenship as an emerging field of study is well established by now (e.g. Cao, 2015; Dobson, 2004; Dobson & Bell, 2006; Latta & Wittman, 2012; Smith & Pangsapa, 2008), but there is a dearth of research on environmental citizenship in the Indian Ocean context. This volume aims to improve our understanding of the intersections between environment and citizenship and showcase their manifestations across this vast region, from Mauritius to Malaysia. The research presented here is informed by the desire to understand and the need to address the environmental challenges facing the peoples of the Indian Ocean. We hope the articles included in this edited collection provide practical insights to improve resilience and adaptation, as well as useful conceptual insights into the nature of environmental citizenship discourse and practice in the Indian Ocean region.

* * * * *

The volume opens with an article by Victor Savage and Lin Qi Feng that sets the context for this special edition. Their piece calls for a profound transformation – a metamorphosis – to tackle the environmental predicament of the Indian Ocean region. Their climate change adaptation program contains five suggestions to survive and thrive under climate change: enhancing regional cooperation; integrating indigenous traditions and folk science; embracing the region's terraqueous geography; creating inland developmental corridors; and ensuring sustainable urban development. Savage and Lin advocate a new paradigm of human-nature relations, informed by indigenous knowledges and based on a terraqueous geography of water, soils and wetlands – one that can generate and sustain new productive activities and facilitate diasporic movements throughout the region. Theirs is in many ways a radical and ambitious proposition, but one that is well suited to a time of existential challenges. Moreover, their concept of terraqueous geography could help us rethink the spatial dimension of environmental citizenship within and beyond the Indian Ocean region.

The article by Erika Techera shifts the focus to environmental stewardship and fisheries governance. Her contribution is also rich in suggestions, derived from a comparative analysis of regulatory frameworks of small-scale fisheries in six Indian Ocean states: three littoral states (Tanzania, Kenya, and Bangladesh) and three island states (Mauritius, Seychelles, and Sri Lanka). Her starting point is the notion that inclusive governance (e.g. civil society engagement and citizen participation in institutions, laws and processes) is essential to address the ongoing challenge of balancing conservation and sustainable fishing. Techera examines the extent to which national fisheries legislation in those states supports public participation in fisheries governance, looking at its three elements: decision-making (i.e. public participation itself), access to information, and access to justice. The article is only a foundational study, but a very useful and important one at that, not least because it provides a sound methodology that can inspire and drive further research; but also because her analysis draws out best practice provisions that can serve to advance fisheries governance in the region, and beyond. To the extent that fisheries governance through public participation reflects the integration of citizens into environmental management, this line of research can also advance environmental citizenship in the Indian Ocean region.

Public participation is also at the heart of Rita Rambaree's contribution. Her article presents a case study of company engagement with civil society in coastal communities in Mauritius. Rambaree works with concepts and realities that are not typically presented together: ecosocial work and corporate social responsibility. Her article shifts the focus of the volume to environmental justice, drawing heavily on Karen Bell's (2014) work on this field to explore the potential of delivering ecosocial work through corporate social responsibility. Rambaree examines the extent to which corporate social responsibility projects in Mauritius deliver on the three dimensions of environmental justice: distributive, substantive, and procedural. Her findings suggest that such projects can have positive social and environmental impacts, but these seem to be limited to employment (through training) and advocacy (through capacity building). However, where others might emphasize limitations and view corporate initiatives as exercises in greenwashing, Rambaree highlights possibilities and advocates for a more collaborative approach to corporate social responsibility. This opens up an intriguing set of questions (and line of research) about the role of the private sector in articulations of environmental citizenship in the Indian Ocean region – not least because the environmental dimension of corporate social responsibility is driving companies to embrace the language of environmental citizenship (see Cao, 2015, pp. 181–188). Yet, as Rambaree points out, inequalities and substantial issues are difficult to address through corporate social responsibility. Indeed, as one of her interviewees notes: looking after the community is ultimately the state's duty.

The state and political elites are at the heart of Ramya Tella's contribution. Her article revisits our environmental predicament – this time in terms of the Anthropocene. Tella explains how this new (albeit contested) geological epoch is forcing us to rethink the law. Drawing upon an intricate articulation of theoretical perspectives to present a discussion of law and climate change through the voices of Indian elites, the article shows how Indian climate elites use the Indian Constitution to underwrite narratives of Indianness. These elites, whilst appreciating the need for a cosmopolitan approach to climate change (framed as a planetary crisis), reassert the nation-state as the paramount political community, and deploy exclusionary forms of identity to address climate change in India. This nationalist construction converges with articulations of postcolonial modernity – only that rather than puncture the nationalist narrative, these articulations serve to negate other sub-national voices (activists and refugees). In other words, activists who challenge the environmental credentials of the Indian state, and displaced peoples from climate breakdown (climate refugees), become the victims of an official environmental citizenship discourse underpinned by an exclusionary conception of Indianness. Tella's article not only provides a critical engagement with environmental citizenship but – alongside the papers by Simpson and Rahman – it also brings the concept to the forefront of this special edition of JIOR.

The critical engagement with the concept of environmental citizenship is also central to Ian Simpson's contribution. His article takes as critical look at the way in which environmental citizenship is deployed as a political instrument in the Arab Gulf states, particularly in the United Arab Emirates (UAE), revealing the contradictions of environmental citizenship in the Arabian littoral. Simpson examines the ways in which labor relations and discourses of environmental citizenship enable or foreclose environmental reform in the Arab Gulf states. His work shows how the citizen-foreigner distinction structures the allocation of rights and duties – although this does not translate into substantial environmental rights afforded to citizens. Instead, the state uses the discourse of environmental citizenship to promote a neoliberal conception of environmental citizenship, one that treats citizens (and non-citizens) as economic units. In other words, environmental citizenship is used to promote a discourse of individual responsibility that manufactures green consumers as the ideal environmental citizens (see Cao, 2017). His analysis also reveals how national identity, in the form of cultural heritage, limits the transformative potential of environmental citizenship by naturalising exploitative relations between citizens (consumers) and foreigners (producers), and marginalizing trans-regional social intermingling in the Arabian littoral. If Tella's and Simpson's articles are anything to go by, there is a long road ahead before the region is ready to embrace the terraqueous geography advocated by Savage and Lin.

The volume ends with an uplifting article that traces the development of Kelab Alami, a community organization established in Malaysia in 2008. Their story is told and examined by one of its founders, Serina Rahman. Her piece has a distinctive personal voice, and is grounded on the author's experience and knowledge of Kelab Alami. Rahman recounts how Kelab Alami evolved from an environmental education club for the local youth to an organization that uses environmental citizenship as the basis for engagement with business, government agencies, and other stakeholders. In its first phase, the youth were trained as citizen scientists, habitat monitors, and community researchers, relying to an important degree on local and traditional ecological knowledge. In a second phase, the youth were trained as ecotourism guides, providing a source of income, and expanding their impact by sharing their understanding of the local natural environment with others. Kelab Alami's engagement with authorities and developers during the second phase echoes the calls for collaborative approaches in Rambaree's article. Notwithstanding the challenges ahead, this is ultimately a story of empowerment that offers a model of action which other communities might be able to adapt and adopt. In essence, Kelab Alami is a wonderful grassroots, community-driven, illustration of environmental citizenship at work in the Indian Ocean region.

*****

Environmental citizenship is an idea whose time has come, and there is no question that our environmental predicament will shape what it means to be a citizen in the twenty-first century (Cao, 2013). This collection of articles showcases the complex field of environmental citizenship, alongside closely related concepts, such as environmental governance, environmental education, environmental justice, and corporate social responsibility. The volume provides a rich, diverse and multidimensional portrait of environmental citizenship in the Indian Ocean context. The articles explore all major geographical dimensions of citizenship: global (Savage and Lin); regional (Savage and Lin, and Simpson); national (Techera, Rambaree, Tella, and Simpson); and local (Rahman). They also explore hybrid political spaces: transnational and trans-regional (Simpson), and the terraqueous geography articulated by Savage and Lin. The explored territories extend across the region, including island states (Mauritius, Seychelles, and Sri Lanka), and littoral states in Africa (Tanzania, and Kenya), Asia (India, Bangladesh, and Malaysia), and the Arab Gulf (UAE). The articles examine the three main actors of environmental citizenship: states (Techera, Tella, and Simpson); citizens (Savage and Lin, Simpson, and Rahman); and corporations (Rambaree). Last but not least, they explore the three main elements of citizenship: membership, understood in terms of status, identity and/or political community (Savage and Lin, Tella, Simpson, and Rahman); rights (Techera, Rambaree, and Simpson); and duties (Rambaree, and Simpson). The dominant themes across the volume are public participation and civil society engagement. These were the intended focus of this special issue all along, but they are also themes that reflect a broader push to articulate citizenship not as passive status but as active practice.

The call for papers proposed a comprehensive list of topics associated with environmental citizenship, including different actors, elements, dimensions, and articulations, many of which are explored in this volume: environmental activism (Rahman); environmental governance (Techera, and Simpson); constitutional texts (Tella); indigenous citizenship (Savage and Lin); corporate citizenship (Rambaree); and environmental pedagogies (Rahman). Yet this collection of articles has only scratched the surface of what is happening in the region, and thus of what can be explored through the lens of environmental citizenship. Indeed, this volume leaves many important issues untouched, or only briefly touched upon, such as: climate refugees (briefly explored by Tella); food citizenship (indirectly explored by Techera); and the rights of nature (evoked by Savage and Lin). I hope this special edition of JIOR has gone some way to enhance our collective knowledge and inspire further research on the discursive and practical articulations of environmental citizenship in the Indian Ocean region.

## Acknowledgements

I would like to thank the Chief Editor, Prof Tim Doyle, for entrusting me with the edition of this special issue, and the Commissioning Editor, Dr Adela Alfonsi, for her invaluable assistance throughout the whole process. I would also like to thank the authors for their contributions and their diligent engagement at the different stages of this project. Last but not least, I would like to extend a special thank you to the peer reviewers. Their generous, substantial and constructive feedback has enhanced each and every paper, and consequently this special edition of JIOR.


## Disclaimer

The contents of this journal are based exclusively on the views of the authors and in no way do these views reflect the views of the IORG Inc., IORAG, IORA, or its member states.


## Funding

Specific funding for this issue has been generously provided by the Australian Government's Department of Foreign Affairs and Trade.

## Disclosure statement

No potential conflict of interest was reported by the author.


## References


Bell, K. (2014). *Achieving environmental justice: A cross-national analysis*. Bristol: Policy Press.

Cao, B. (2013). Environment and citizenship: Rethinking what It means to Be a citizen in the 21st century. In J. Arvanitakis & I. Matthews (Eds.), *The citizen in the 21st century* (pp. 155–166). U.K.: Inter-Disciplinary Press.

Cao, B. (2015). *Environment and citizenship*. London: Routledge.

Cao, B. (2017). Consuming environmental citizenship, or the production of neoliberal green citizens. In J. Louth & M. Potter (Eds.), *Edges of identity: The production of neoliberal subjectivities* (pp. 25–51). Chester: Chester University Press.

Dobson, A. (2004). *Citizenship and the environment*. Oxford: Oxford University Press.

Dobson, A., & Bell, D. (Eds.). (2006). *Environmental citizenship*. Cambridge, MA: The MIT Press.

Latta, A., & Wittman, H. (Eds.). (2012). *Environment and citizenship in Latin America: Natures, subject and struggles*. New York: Berghahn.

Smith, M., & Pangsapa, P. (2008). *Environment and citizenship: Integrating justice, responsibility and civic engagement*. London: Zed Books.

# Climate change adaptation: the need for an Indian Ocean regional metamorphosis

Victor R. Savage and Lin Qi Feng

**ABSTRACT**

The Indian Ocean contains one of the busiest trade routes in history, connecting East Asia, Southeast Asia, South Asia, the Middle East, and East Africa, as well as giving passage to Europe. Given its historical and economic connections and diffusion of ideas, religions, and cultural traditions, the Indian Ocean region's (IOR) fertile cultural complex can serve as an important platform for addressing climate change. In order to survive and thrive under climate change, societies in the IOR need to undergo profound transformation or metamorphosis. We briefly discuss the history and major climate change issues of the IOR before proposing five suggestions: engaging in regional cooperation on climate change; combining the region's folk science and indigenous traditions with modern science and technology in the formulation and implementation of climate strategies; embracing the region's terraqueous geography; initiating development corridors inland; and ensuring that cities and urban development are sustainable and liveable.

## Introduction

The increasing mantras on climate change reflect three themes. First, there is growing international concurrence of the evident role and severity of human impacts on climate change through varied activities ranging from fossil fuel use, industrialization, and landscape changes. The second theme is the resulting imperative of the first: we need to shift away from 'business as usual' trajectory – growth-based economics driven by capital accumulation, consumption, and powered by fossil fuels. Third, the tropical region is expected to undergo immense changes, more so than the temperate region, making countries in the tropics vulnerable to climate change. The Indian novelist Amitav Ghosh describes our present climate predicament, which was borne out of modernity's disregard for the environment and obsession with freedom, and perpetuated through capitalism, imperialism, and fossil fuel use as 'The Great Derangement' (Ghosh, 2016).

We use metamorphosis to describe the profound transformation which societies in the Indian Ocean region (IOR) need to undergo in order to survive and thrive under climate change. In this paper, we briefly discuss the history of the IOR and major climate change issues in the region. We then provide suggestions which hopefully will lead us towards the desired metamorphosis. While our focus is on climate change adaptation, we believe our suggestions are also pertinent to mitigation and that both adaptation and mitigation should be pursued in tandem in the IOR.

## History of the Indian Ocean region

William Kirk (1951) was the first academic to outline the development of an 'Indian Ocean Community,' a diverse grouping of states. Historically the states, or their predecessors, have traded and communicated with one another under relatively peaceful conditions. More recently, the states have been collectively referred to as the Indian Ocean region (IOR). In terms of geographical terminology, the IOR is not a 'formal' region in the sense of a shared history, culture, or social systems, but is best seen as a 'functional' region that is bound together by similar functional interests, such as shipping, fishing, and environmental issues.

The former kingdoms, thalassic empires, and ports around the Indian Ocean thrived on creating a terraqueous history (Bashford, 2017). While the Eurasian continent joins Asia and Europe, the Indian Ocean, as the 'Afrasian Sea' (Ray & Alpers, 2007), links Africa and Asia. Despite its far-flung spatial extent across a diversity of vernacular cultures, three pan-Oceanic cultural and migratory movements have linked the IOR over the centuries. These movements can be seen as a rudimentary basis for considering the IOR as a 'formal' region that has been shaped by historical processes and cultural exchanges.

First, the Austronesian migration across the Indian Ocean has linked the Southeast Asian peoples from the Malay world (*Alam Melayu*) with the Malagasy people and language that is native to Madagascar (Bellwood, Fox, & Tryon, 2006). The cultural links are strong and show that the Malayo-Polynesian language is also the main language in Malagasy (Linton, 1955).

Second, Indianization has been a powerful cultural force through which Hinduism and Buddhism were spread and has linked Southeast Asian states, polities, and cultures to the Indian sub-continent. The most pervasive aspect of Indianization has been Sheldon Pollock's (2006) thesis on the spread of Sanskritization across India and Southeast Asia. The 'Sanskrit cosmopolis' transcended multilingual and multiethnic populations in India and Southeast Asia, expanding across ecological, cultural, and political boundaries. Sanskrit had spread almost by 'cultural osmosis' because of its expressive power in communication, its translocality, its transethnicity, and the stability and dignity that is conferred by its grammar (Pollock, 2006, pp. 254–256). The spread of Buddhism from India created important linkages between India, Sri Lanka, and East Asia. Chinese traders, Buddhist monks, and pilgrims provide many accounts of travels from East Asia to Southeast Asia and India (Haw, 2017).

Third, Islam has been a powerful religious force in the IOR, linking East African peoples with South Asia and Southeast Asia. Using the established trading networks, Muslim traders spread their religion in coastal East Africa, South Asia, and Southeast Asia. The Muslim Mogul reign in India has become an integral aspect of Indian architecture, foods, culture, music, and dressing. In Southeast Asia, Islam became an anchor in the kingdoms and states of the Malay world, spreading from Aceh, Melaka, and *pesisir* or coastal towns in Java, to Sulawesi, Brunei, and the Sulu Archipelago.

This historical crisscrossing of cultures, religions, trading produce, and ideas results in a fertile cultural complex. For example, the original indigenous spirits of seafaring communities in Southeast Asia were later joined, and in some cases supplanted, by the localization of the 'macrocosmic' pantheon of gods and deities from Islam, Hinduism, Christianity, and China (Andaya, 2017). Vernacular travelogues that emerged alongside steam travel in the Indian Ocean during the nineteenth century showed that a vernacular intellectual milieu existed that was not captured by European intellectual 'hegemony' (Green, 2018). In the present time, this cultural complex can be a source of indigenous adaptive systems as well as a platform for dialogue and collaboration amongst states and communities over environmental and climate change issues, and possibly be supplemented in this endeavor by the experience of developed countries.

## The impact of climate change in the Indian Ocean region

According to the Fifth Assessment Report by the Intergovernmental Panel on Climate Change (IPCC), 'it is extremely likely that human influence has been the dominant cause of the observed warming since the mid-twentieth century' (IPCC, 2014a, p. 17). In an October 2018 special report, the IPCC noted with high confidence that 'global warming is likely to reach 1.5°C between 2030 and 2052 if it continues to increase at the current rate' (IPCC, 2018, p. 4). There are at least three areas through which climate change and El Niño – the warm phase of the El Niño–Southern Oscillation (ENSO) – will impact states and peoples in the Indian Ocean region.

### *Warming of the Indian Ocean*

Recent studies indicate that the Indian Ocean is the warmest ocean and that the sea surface temperatures (SST) are greater than 28°C (Luo, Sasaki, & Masumoto, 2012). From 1950 to 2016, the surface of the Indian Ocean basin has warmed by 0.11°C, compared to 0.07°C and 0.05°C for the Atlantic and Pacific Oceans respectively (IPCC, 2014c, ch. 30).

Studies demonstrate more links between the Indian and Pacific Oceans with impacts on terrestrial areas still subject to scientific enquiry. The western Indian Ocean between 1901 and 2012 shows evidence of SST warming and is related to the Pacific La Niña events (Roxy, Ritika, Terray, & Masson, 2014). While globally oceans account for 93 percent of all heat accumulated since 1970 due to human greenhouse gas emissions, the Indian Ocean alone absorbs 70 percent of all heat taken up by global oceans (Lee et al., 2015). The rise in the warming of the Indian Ocean could have a modulating impact on the Pacific Ocean in the twentieth and twenty-first centuries (Luo et al., 2012). The question is whether the Indian Ocean influence will have an impact on the El Niño-La Niña cycle globally. Unlike the Pacific Ocean, on which long term research has been conducted, the Indian Ocean has been relatively understudied. Only in 2008 was an Indian Ocean multilateral collaboration established called the Research Moored Array for African-Asian-Australian Monsoon Analysis and Prediction, or RAMA. RAMA is investigating the variability of the Indian Ocean Dipole (IOD) – an irregular oscillation of SST in the Indian Ocean – in terms of east–west anomalies of precipitation and drought as well as its societal consequences (McPhaden et al., 2009).

The increase in ocean temperatures is already affecting coral reefs. Some two-thirds of the Indian Ocean's 12,070 $km^2$ of corals are already destroyed or critically damaged (Michel, 2012, p. 115). Climate change is putting at risk the biodiversity of the region and inducing species extinction. Elizabeth Kolbert's (2014) *The Sixth Extinction* is a chilling description of what might take place in the following decades. In the five previous mass extinctions, 80 percent of all organisms were exterminated. The imminent species extinction might include Homo sapiens (Imanishi, 2002; Rees, 2003).

### *Sea level rise*

One of the most worrying outcomes of climate change for littoral states and coastal cities globally is the imminent threat of sea level rise. Thermal expansion of sea water and irreversible melting of ice caps will result in sea level rise. The IPCC estimated that between 1901 and 1990 the rate of global mean sea level rise was 1.4 mm per year, and between 2006 and 2015 the rate was about 2.5 times higher at an unprecedented 3.6 mm per year (IPCC, 2019, pp. SPM-10). Global mean sea level rise for the period 2081–2100, compared to 1986–2005, is projected to range from 0.26–0.92 m, depending on the representative concentration pathways (IPCC, 2019, pp. SPM-23). Sea level rise will cause the frequency of extreme sea level events at most locations to increase and collapse of Antarctic ice sheets could increase Antarctica's contribution to sea level rise (IPCC, 2019, pp. SPM-23).

The scientific predictions and forecasts are not very positive for the IOR. Sea level rise is not uniform in Indian Ocean coasts. The coasts of the north Indian Ocean are said to have risen 12.9 cm per century (Han et al., 2010, p. 546). Specifically, due to the Indo-Pacific warm pool effects, certain mid-ocean islands like the Mascarenhas archipelago, coasts of Sumatra in Indonesia, and the north Indian Ocean may experience significantly more sea level rise than the global mean (Han et al., 2010, p. 549). Conversely, the Seychelles islands and the east coasts of Kenya and Tanzania may witness little or no sea level rise (Han et al., 2010, p. 549). While eastern Pacific Oceanic rise in temperature had impacts on the IOR thousands of kilometers away in Southeast Asia (e.g. droughts, forest fires, haze), IOR governments need to brace themselves for new scientific reports that temperatures and sea-levels in the Pacific and Indian Oceans are linked, which is resulting in a rise in temperatures and sea levels in the Indian Ocean (Han et al., 2010; Luo et al., 2012; Mochizuki, Kimoto, Watanabe, Chikamoto, & Ishii, 2016).

Many of the Indian Ocean states have major ports, capitals, and development zones abutting the coastlines that are vulnerable to sea level rise and related factors, such as extreme weather events. States that are at risk include Australia, those in Southeast Asia, India, Bangladesh, and Sri Lanka in South Asia, Yemen and the Gulf States in the Middle East, and Somalia, Kenya, and Mozambique in East Africa. The island states of the Maldives and Seychelles are likely to be hit hardest within the next couple of decades. Studies show that sea levels have risen since 1960 along all Indian Ocean coasts, except Zanzibar (Han et al., 2010, p. 546). Bangladesh is especially vulnerable: in 2000, about 40 percent of its land is in low elevation coastal zones where 46 percent of its population lives (McGranahan, Balk, & Anderson, 2007). Agriculture in its coastal zones has been and will continue to be affected by multiple climate change factors (Huq, Hugé, Boon, & Gain, 2015; Ruane et al., 2013).

### *Droughts*

While El Niño is associated with reduced precipitation, climate change will likely induce extreme weather events, leading to floods, droughts, and irregular weather patterns. Combined with increasing populations, growing economic activity, and reliance in agriculture on irrigation, climate change will increase the water stress of the IOR (IPCC, 2014c, pp. 1334–1338). Shortage of freshwater, as is already taking place in the Sahelian area, will affect societies in two related ways. First, it will exacerbate public health issues and more generally affect a country's development. Second, it will lead to lower agricultural production in East Africa, Myanmar, and India, which could lead to famine and malnutrition. According to *The State of Food Security and Nutrition in the World 2017*, about two thirds of the world's undernourished people and severely food-insecure people are in the IOR. The estimated number of undernourished people in 2014–2016 in Eastern Africa, South Asia, and Southeast Asia are 125.8, 271.6, and 64.8 million, respectively; this constitutes almost two thirds of the estimated total of 789.1 million. The estimated number of severely food-insecure people in 2014–2016 in Eastern Africa, South Asia, and Southeast Asia are 126.9, 235.9, and 45.0 million, respectively; this constitutes more than two thirds of the estimated total of 666.5 million (FAO, IFAD, UNICEF, WFP, & WHO, 2017, pp. 86–89). Ecosystems, which provide a range of benefits for human society, will be severely damaged without a constant supply of freshwater.

Climate change is already having a major impact both on land and in the seas. Droughts induced by El Niño occur periodically in Indonesia (Kalimantan and Sumatra) and mainland Southeast Asia, and often cause massive forest fires that result in palls of haze in Malaysia, Singapore, Brunei, and southern Thailand (Glover & Jessup, 1999). Heat waves during the summer of 2018 have caused human deaths, forest fires, and affected economic activities (Chow, 2018).

## The limitations of western human-nature thought

The counterpoint to the climate-induced threats facing humanity is the geophysical impact of industrial human society, which is estimated to result in climate change for millennia to come (Archer, 2009). Our climate predicament has revealed what is truly universal in human society: our biophysical existence in the planetary commons of the atmosphere. Yet the origin of our climate crisis and broader environmental predicament arose unevenly out of human society. Historically the developed countries are responsible for the majority of greenhouse gases, though developing countries, including those in the IOR, are catching up to developed countries in contributing to climate change. A study estimated that the total relative contribution to climate change from 1850 to 2005 is 53–61 percent from developed countries and 39–47 percent from developing countries; increasing greenhouse gas emissions from recent years from developing countries do not significantly alter this relative contribution (Wei et al., 2016, pp. 641–642).

The indigenous cultural complex of the IOR can be a rich source of ideas not only for climate adaptation but also climate mitigation. Recognizing that the link between principles and beliefs, and actions can be very tenuous (Passmore, 1980, p. ix), and that ideas circulate between the Western sphere and the rest of the world, this indigenous cultural complex also serves as a contrast to Western ideas of the human-nature relationship. In addition, the cause–effect and object-subject polarities (e.g. life-death, generation-extinction, and appearance-disappearance) endemic in Western philosophies are less pronounced in Asian philosophies and religion where the emphasis is on Great Harmony (Dao), self-generating wholeness, or holism (Chinese philosophy), the emancipation of birth–death (nirvana) and the belief in relational nature (Buddhism), and the unity of the cosmos and humanity (Hinduism) (Dallmayr, 2011, pp. 141–153).

Some critics have traced the fraught environmental relationship of the West to the doctrines of the Judeo-Christian religion, the classic example being Lynn White's essay (1967). In *Man's Responsibility for Nature*, John Passmore gave a careful account of the complexity and diversity of traditions of the West (1980). Passmore pointed out two critical concepts therein. First, it is clear from the Genesis that man has dominion over beasts, which could be interpreted as man being an absolute ruler who cares for his dominion insofar as it profits him. Second, nature is not sacred, which leaves man uninhibited in his use of it. Passmore summarized the special attitudes that Christianity (not Judaism) encouraged towards nature:

> that it exists primarily as a resource rather than as something to be contemplated with enjoyment, that man has the *right* to use it as he will, that it is not sacred, that man's relationships with it are not governed by moral principles. (30)

To these attitudes were added the Stoic view that man's dominion over nature is based on his rationality, and Baconian and Cartesian humanism, ultimately resulting in man's duty to transform the world through science and technology into a second Garden of Eden (12–24).

In the shadow of this dominant perspective are two important minority traditions in Western civilization (28–40). First, man is a steward who looks after God's creation. In his 2015 encyclical *Laudato si'*, Pope Francis argued that humans' 'dominion' should be understood more properly in the sense of stewardship and invoked the Eucharist as another source for human stewardship (Pope Francis, 2015, lines 116, 236). Second, man as co-operating with nature in an attempt to improve it, thereby leaving to his descendants a more fertile nature. Alternative Western traditions can also be found in Paganism and Romanticism.

The dominant Western perspective of nature has been very influential from the seventeenth century onwards. One of the impediments in dealing with contemporary climate change challenges is the way through which Western civilizational influences (from the *mission civilisatrice*) and modernized thinking (such as technology, science, and rational thought in general) have shaped global responses of climate mitigation and adaptation. After over 450 years of Western colonial influence in the 36 states of the IOR it has been difficult for political leaders, bureaucrats, and governments to wean themselves from Western political ideologies, modernized responses, technological solutions, and Western thought in confronting environmental problems.

The colonial experiment which was implemented for a century might have opened the doors to science, technology, and capitalism, but it has also left a legacy of major developmental failures in many IOR countries. Where the colonial experiment was implemented under draconian methods (the culture system) as in Indonesia under the Dutch, the failure was already evident in the nineteenth century in Java. Dutch obsession with mercantilist trade, profits, cash crops, and plantation agriculture led to massive famines in Java by replacing food with cash crops, encouraging population growth that changed the human-land ratio to unsustainable levels (Geertz, 1963), and introduced capitalism which undermined the 'subsistence agriculture,' replaced the *'adat'* land tenure system, corrupted the safety-first moral economy, and exacerbated rural inequalities (Scott, 1976). Agriculture became a victim of a capitalist sponsored drug addiction: crops were nursed on heavy doses of chemical fertilizers and pesticides, which became a massive ecological problem for Indonesia (Oka, 1991; Pimentel, 1996).

South Asian states such as India, Pakistan, Sri Lanka, and Bangladesh have had similar colonial experiences which have undermined indigenous philosophies and relationships with their environment and denigrated indigenous and traditional human-nature systems in favor of modern technological systems. As Pankaj Mishra (2012, p. 45) argues in *From the Ruins of Empire*, the West's subordination of Asia was not only political and military but more important it was a kind of conquest which was intellectual, moral, and spiritual. According to Edward Said (1978), the Orientalism fostered by the West is a policy of mental colonialism of the local population. Under the umbrella of modernization, the West continued to control developing Asia through technology, science, and industrialization. The tragedy is that indigenous thinking, folk science, and traditional methods of human-nature relationships were discarded in the dustbin as old fashioned, backward thought and behavior. In the case of Singapore, the government adopted a possibilistic environmental ideology but adopted it to suit the local cultural system and the confined spatial constraints (Savage, 2004).

## Achieving metamorphosis

In terms of mitigation, the 2018 IPCC special report noted that pathways that limit global warming to 1.5°C 'would require rapid and far-reaching transitions in energy, land, urban, and infrastructure (including transport and buildings), and industrial systems' which are 'unprecedented in terms of scale, but not necessarily in terms of speed' (IPCC, 2018, pp. 4, 21). The adaptation measures needed in countries that are most vulnerable to climate change are equally unprecedented as the strains due to growth in population and economic activity are felt keenly. The US military has identified four areas – sub-Saharan Africa, the Asia-Pacific region, the Arctic, and the Middle East – where climate change will cause national unrest and political instability (Holland, 2016). In the IOR, non-traditional security threats, such as those arising from environmental and climate issues, are as important as traditional security threats (Chellaney, 2010). To minimize such threats, active pursuit of climate change mitigation and adaptation is needed.

Social and political changes tend to be painful and difficult to face. For many small states and cities in the IOR, the issue of climate change needs to be accepted as a given that must be embraced and integrated in the social, cultural, and political fabric of their societies. In other words, climate change adaptations require a paradigm shift in the thinking of political leaders in the IOR and indeed all over the world. Adaptation is context specific. This means adversity must be turned into advantage, environmental determinism must be recast into eco-possibilistic thinking (Berdoulay, 1974), and change must be converted into metamorphosis. The metamorphosis of adaptation must manage two areas: hardware adaptation (e.g. infrastructure, urban scaffoldings, and technical issues) and software adaptation (e.g. education, technical training, and folk culture understanding).

Metamorphosis calls for devising measures that consider the unique cultural circumstances of the IOR. We highlight the importance of regional cooperation in mitigating and adapting to climate change and four broad areas that states need to focus their national energies and limited resources on. First, embrace indigenous adaptations and folk science to complement other organizational, technical innovations, and ecological conditions. Second, as littoral states of the Indian Ocean, recognize the importance of their terraqueous geography and leverage on trade and exchanges. Third, initiate development corridors that lead away from threatened coastal areas. Fourth, given that a large percentage of populations in the IOR reside in urban areas, nucleated settlements (towns and cities) must be given special attention in climate change programs.

### *Regional cooperation*

Regional cooperation plays an important role in the climate response of the IOR. Since it is difficult to find global common solutions to climate change, as the Paris Agreement has shown, there is greater incentive to discuss, debate, and dialogue climate change measures at regional, national, and local levels. An example of regional efforts to address common interests is the Mediterranean Action Plan, which was created in 1975 as the first United Nations Environmental Programme initiative to be developed under the latter's Regional Seas Programme. As a result, the Mediterranean states have been able to come together to discuss environmental solutions with varying degrees of success (Frantzi, Carter, & Lovett, 2009; Skjærseth, 2002), despite great economic, cultural and political differences between them.

The Indian Ocean can emulate in part what the Mediterranean states are doing. The prima facie challenges are establishing an IOR-wide collaboration on climate change, either through the Indian Ocean Rim Association (IORA) or through a dedicated climate change mechanism (Sembiring, 2018) and focusing on the important but at times seemingly unurgent climate adaptation. While regional cooperation is not without its problems and ecological security has not been a main focus, prior experience of cooperation in 'sustainable development' and more recently sustainable development goals, and common experience such as the devastating Indian Ocean earthquake and tsunami of 26 December 2004, can serve as a source of commonality.

There is immense potential for regional cooperation on climate adaptation in the IOR. The Indian Ocean states share three things in common. First, all of them are vulnerable to the ravages and challenges of climate change and hence they can benefit by exchanging ideas and methods of responding to this major environmental predicament. Second, all the states share the Indian Ocean geographical entity and hence have something in common to begin a dialogue. One common outcome of global warming is sea level rise and this will affect many littoral states along the Ocean. Third, the Indian Ocean covers three continents, Oceania, Asia, and Africa, which despite large differences have historically been active in trade, religious diffusion, cultural exchanges, and human migration of peoples. Given the vastly different ecosystems, the exchange of ideas, goods, and cultural ways of living will be of great benefit to communities in the region in tackling global environmental challenges like climate change. To enhance regional cooperation on climate change adaptation, the Indian Ocean states need to develop an intra-IOR platform to address common problems and future challenges, and find commonalities in climate change responses which will affect many socio-economic issues.

In order to have a meaningful and productive engagement amongst Indian Ocean states, several climate change commonalities need to be identified. First, states need to identify the common climate change issues they are facing so that a dialogue can take place. Second, they need to prioritize their most pressing climate change challenges. This could be sea level rise, water shortages, or agricultural problems arising from heat waves, flooding, or dwindling precipitation. Third, there must be a bold admittance by governments that they lack expertise to solve problems so that exchanges in expertise can be made from within the region. Political leaders, academics, and bureaucrats must be wise enough to realize that not all modern engineering technological solutions (e.g. geoengineering) are the panacea for climate change woes.

## ***Indigenous traditions and folk science***

Indian Ocean states need to shed their colonial mentality. The Orientalist thinking identified by Said (1978, 1993) needs to be reversed. The challenges of climate and environmental change require a metamorphosis in thinking, i.e. integrating traditional views with new ideas to meet the challenge of our present sustainability crisis (Duara, 2014). Pankaj Mishra (2012) argues the Asian renaissance can only take place with a revolt against Western thought. The Indian Ocean states would require new ways of thinking to deal with current and future problems arising from climate change.

Against the Western superior, top-down, unequal human-nature relationship, Asian and indigenous nature relationships are holistic, recognize equality of living denizens, and are respectful of all organisms of earth. Anna Peterson (2001) provides an interesting view of how Asian religions and indigenous peoples perceive the world around them and interact with nature through 'relational' perspectives rather than a top-down, hierarchical nature relationship.

Further, indigenous people's system of working with nature could be described as a 'folk science' that has evolved over long periods of adaptation and experience. Many indigenous methods and practices have been successfully applied for millennia as adaptive measures (e.g. the Balinese *subak* irrigation system [Lansing, 1991]) to stem weather variations and these should not be discounted in current deliberations. This folk science is tied to the peoples' folk lore, which possesses its own system, as Bronislaw Malinowski (1922) had shown for the people of the Trobriand Islands. Although folk science is a contested term (Berkes, 1993; Bicker, Ellen, & Parkes, 2003), it can be combined with modern science to solve current problems (Agrawal, 1995; Briggs, 2005; Mercer, Dominey-Howes, Kelman, & Lloyd, 2007). Patrick Kirch (2010) argues that there can be better methodological rigor to studies in history and the social sciences. Studies of Pacific island communities and in this case the Indian Ocean need to discriminate between cultural traits (and indigenous technologies) that are shared (*homologies*), derived or innovated (*analogies*), and borrowed (*synologies*).

From Africa to Australia and across land masses and islands in the Indian Ocean there are many indigenous peoples who have survived thousands of years living in various ecosystems. If there is one cardinal rule of survival humans need to adopt, it is that we do not subscribe to a one size fits all system of sustainability. Jared Diamond (2012) demonstrates how sustainability of indigenous peoples in the Pacific and New Guinea is predicated on a human-land carrying capacity equilibrium, adapting to the hardships in nature and exploiting carefully natural resources over varied ecosystems. James Scott (2009) showed how swidden tribal groups in the area of *zomia* (eastern India, southern China, and mainland Southeast Asia) led autonomous, independent sustainable lives for thousands of years in their highland ecosystems.

An important contributing factor to climate change is the capitalistic economic system which emphasizes exponential economic growth, resulting in increasing human appropriation of natural materials and, consequently, ecological destruction (Foster, Clark, & York, 2010) and social inequity (Piketty, 2014). The critical question is whether the Indian Ocean states can tame capitalism and reorganize its economic systems to provide goods and services in a more sustainable and equitable way. David Harvey has provided much food for thought on the inelasticity and elasticity of capitalism in this regard (Castree & Gregory, 2006).

The historical records of the IOR provide a pessimistic picture of the sustainability of kingdoms, cities, and civilizations. Ironically, one variable in the expansion and implosion of civilizations has to do with climate change. Victor Lieberman (2003–2009) provided a comprehensive documentation of the expansion and collapse of civilizations in the Eurasian landmass, which was later updated with tree-ring data (Lieberman & Buckley, 2012). He found 'strange parallels' between the expansion and contraction of kingdoms and empires in Europe, South Asia, East Asia, and Southeast Asia over several centuries arising not from global warming but from the El Niño La Niña Southern Oscillations (ENSO). It seemed evident that tribal groups and peasants in *zomia* understood the precarious weather conditions arising from prolonged drought and hence developed a whole culture of adaptation towards water scarcity. In mainland Southeast Asia this adaptation included building large reservoirs (*barays*) and an irrigation system in Angkor Thom, villages collecting rain water in huge earthen jars, worshiping the *naga* (underwater serpent) to ensure rains, using portable houses in the *Tonle sap* (Cambodia) to maintain proximity to the changing lake shore during the wet and dry seasons, and partaking in the water festival just before the rainy season to ensure precipitation for the beginning of the rice cycle.

In *Dark Emu*, Bruce Pascoe (2018) presents a startling reinterpretation of Australian Aboriginal society as a complex agricultural civilization, especially in the Murray-Darling Basin, that had evolved over thousands of years. This civilization was deracinated and the land degraded upon the arrival of settler-colonists. This revisionist account reveals the potential of arid lands for agriculture, highlighting the potential value of folk science for climate change adaptation. Cross-environmental studies in the IOR can furnish refreshing ways through which human communities can adapt to growing aridity. These studies can compare indigenous adaptive systems as well as larger macro-scale systems.

Folk science can be supplemented by modern science that follow basic environmental principles, such as solar panels, and eschew overly complex technology such as nuclear energy. People can overcome challenges by adopting possibilistic (humans as arbiter of change and decision-making) thinking, adaptation, and behavior (Savage, 1997). Israel has shown the world it can transform its arid landscape into a harvestable garden of fruits, cereals, and vegetables (Tal, 2007). China, learning from the Israeli experience, is likewise turning the Gobi Desert into an agricultural, industrial, solar energy farming, and residential area (Xie et al., 2018). These examples underscore the importance of metamorphic thinking – the transformative power to capitalize or even reverse adverse environmental situations.

Another example is Singapore's water recycling and desalination efforts to ensure water self-sufficiency (Tortajada, Joshi, & Biswas, 2013). However, such water technology is expensive and energy-intensive and would only be feasible under unique circumstances. Countries in the arid states of the Middle East and East Africa as well as Pakistan should embark on varied methods of water enhancement by taking advantage of the economies of scale for nucleated places such as towns and cities.

### *The Indian Ocean terraqueous geography*

It seems an opportune time for states to embrace a terraqueous geography, turning the Ocean into what Claudius Ptolemy (1932) thought was a lake and develop more intra-Oceanic exchanges. The lake metaphor has endorsed over the centuries with geopolitical connotations: a Roman lake, a Portuguese lake, a Dutch lake, a British lake, and an American lake. Each of these external powers capitalized on the long tradition of movement of goods, services, peoples, and ideas in the IOR. The idea of an Indian lake gave each succeeding power the idea of political control within a closed corridor of trade through control of 'vital sea gates' (Kirk, 1951, p. 163). Hence the Portuguese reign was marked by outposts from Mozambique to Malacca and included Goa, Sri Lanka, and Timor. The Dutch ruled their trading empire in the Indian lake under the Dutch East India Company (often referred to by its Dutch acronym VOC) from South Africa to Indonesia, including India, Sri Lanka, and Malacca. The Indian lake has thus been visited and in some cases colonized by civilizations from without. On the western front came the Europeans and on the eastern the Chinese (Haw, 2017).

The IOR needs to put aside political differences and find commonalities in its varied problems. The interaction of peoples across the Indian Ocean is unmatched compared to the Pacific and Atlantic Oceans. Indeed, the Indian Ocean was a scene of pacific interactions for centuries based on trade, diffusion of religious ideas, and migration of peoples. Different produce across the oceanic ecosystems produced goods for barter trade, developed financial accounting, and created the nascent development of capitalism (Abu-Lughod, 1989; Blaut, 1993). Even when the Roman Empire extended its spatial reign over the Indian Ocean for about five centuries, it did not disrupt the indigenous trade but served as a security umbrella against piracy (Warmington, 1974).

It was the Portuguese that changed the geopolitical equation when they entered the Indian Ocean in the fifteenth century and captured Goa and Malacca in 1511 under its belligerent colonialism. The Portuguese thrived on the stable trading infrastructure that was already established. Alfonso de Albuquerque and the Portuguese apothecary Tome Pires noted the acquisition of Javanese and Arab charts and maps of the Indian Ocean that aided their understanding of the geography of the IOR (Sollewijn Gelpke, 1995). Western colonialism did not create but rather inherited a well-established network of ports, an indigenous flow of trade, and cherry picked the produce desired by their markets. In the unfolding process of decolonization, IOR states should reestablish historical connections amongst themselves to ride the new Afro-Asian Renaissance (Ibrahim, 2010; Okumu, 2002).

While Indian Ocean states are threatened by climate change impacts such as sea level rise and ecosystem degradation, there is an urgent need to extend terrestrial interactions into the vast ocean and seas. Specifically, states need to manage the terraqueous space as routes for travel and transport, as natural resources for marine-based economies or 'blue economies,' as buffers for climate change, and as economic networks for trade and commerce (Llewellyn, English, & Barnwell, 2016). Given increasing globalization, there is a need to remain open to 'transnational' relationships in many areas – trade, migration, tourism, cultural exchanges, political cooperation, security enhancements, educational partnerships, and environmental relationships. Unlike terrestrial geography which is divided up by states, terraqueous geography is still governed in part by the law of the sea (*mare liberum* and *mare clausum*) and freedom of navigation with large expanses open as international waters. A terraqueous geography must move away from fixated land-based thinking and embrace a more fluid world-view where migration and mobility take centre-stage. This world-view helps to undergird diasporic movements, maintain a nimble attitude towards adaptation to environmental challenges, and alert people to imminent changes in their interactions with ecosystems.

The terraqueous geography of the Indian Ocean provides good potential for economic, cultural, and political development. Under the IOR umbrella, there are secured economic and political anchors. Regional groups such as ASEAN and SAARC are important pivots for cooperative endeavors between regional groups. In the future, important economies such as India and Indonesia have the potential to be economic drivers for the IOR. Most of all the Chinese Belt and Road Initiative (BRI) has a maritime route from China through the South China Sea, the Straits of Melaka, and into the Indian Ocean, which promises to invigorate the trade and development of adjacent littoral states. The Chinese aggressive pursuit of the BRI on the one hand might spur development in the IOR, but for geopolitical skeptics the rising dragon's establishment of strategic beachheads in the Indian Ocean is seen yet again as another possible external colonial power creating a Chinese lake and leading to 'a new version of colonialism,' as the Malaysian Prime Minister, Dr Mahathir, articulated during his August 2018 visit to China (Mahathir Warns Against New 'Colonialism' During Visit to China, 2018).

The sub-text of the Chinese BRI in the Indian Ocean is a contentious issue especially for the US, India, and Australia. China for centuries has been a terrestrial country, but in the last 30 years it has increasingly turned its attention to marine domination to safeguard and promote her national interests. As Professor of History from Guangzhou's Jinan University, Milton Nong Ye states: 'China is not safe and has been invaded many times. The way to protect ourselves is to build a great wall of the sea, and you do that with big ships and strong islands' (Hawksley, 2018, p. 39). The Indians have maintained a tacit Monroe doctrine of the Indian Ocean as their geopolitical sphere and thus the aggressive Chinese strategic intrusion into the IOR is a cause of great strategic discomfort for India (Singh, 2018). If the Indian Ocean becomes a place of geopolitical conflict, it could distract and derail all Indian Ocean regional efforts to cooperate together for climate change adaptation and mitigation.

The idea of building walls for defensive purposes, for keeping foreigners out seems an outmoded concept in a globalized and cosmopolitan world. With sea level rise, the walls of the future are likely to be sea walls and environmental bunds. Both Bangladesh and Singapore have shown the way for future ecological protection. The littoral states of the IOR need to discuss and rethink how to safeguard their coastal areas against sea level rises, protection against hurricanes, tsunamis, and flooding from high nip tides.

Indonesians refer to their country as *tanah air* or 'land and water' and aptly defines the new geographical frontier for cooperation and development. This ecological ecotone, environmental fusion, and green synthesis should be a venue of environmental thought for the future. In the IOR, Southeast Asians, with their vast archipelago, have used their terraqueous geography for survival for centuries. In their island world they have plied the seas, exploited mangroves, and developed a trading network along coastal and riverine areas. For centuries the region also had 'sea nomads' or sea gypsies (Sopher, 1977) who lived on boats, living life as mobile entities, and deriving subsistence from both the seas and mangroves. Instead of destroying the rich biodiversity of mangroves through land reclamation and port, transport systems, and infrastructural developments, governments should initiate mangrove protection programs. Mangroves not only provide a source of natural resources which local communities depend on, but are also important coastal corridors of biodiversity preservation and protection buffers against sea erosion, tidal waves, and tsunamis (Friess et al., 2019).

### ***Developmental corridors***

Developing new inland areas to substitute affected coastal areas requires long term planning and development. Governments need to target new corridors of development inland. Such developments will become growth poles and attract migrants affected by sea level rise, spur the national economy, and allow developers to leap frog into more sustainable infrastructural and agricultural developments. In India, new technologies are allowing India to leap frog costly infrastructures in telecommunications. Due to mobile phones India need not invest in expensive land lines (Sivaram, 2017). Similarly, with local solar power India need not construct a national power grid, which means it could easily decentralize power to villages. The revolution in the digital and telecommunication technology will facilitate a widespread democratization of information which will enable the spread of environmental knowledge and information to a wider mass base and give them better opportunities to adapt to ecological challenges.

Rather than accepting the fate of environmental determinism arising from climate change, the tropical states of the IOR can develop ecological creativity. New development zones should pay greater attention to ecological stresses and strains and customize projects to fit into the environment. By developing new corridors of development inland, states will be able to relieve the current pressure of population growth and infrastructural development on coastal areas (e.g. land reclamation, urban extensions, and transport networks).

Specifically, these new developments inland will help to arrest the destruction of mangrove and wetland areas, protect coastal coral reefs, stop pollution (e.g. sewerage, industrial effluents, agricultural chemical run-offs), prevent the dwindling of fisheries, and relieve pressure on coastal area 'dead zones.' Governments need to target a new paradigm of human-nature relations based on a terraqueous geography of water, soils, and wetland areas and look at new productive activities in these ecotone zones. To enhance marine fisheries, political authorities, and corporations could expand aquaculture of marine foods (e.g. fishes, shell fish, prawns, and crabs). The growth and development of aqua-farming will provide more sustainable food sources, protect marine biodiversity, and prevent overfishing of certain popular marine foods.

The relief of port and industrial development in coastal areas will also enhance coastal tourism, a big money earner for many counties such as Indonesia (Bali), Thailand (Phuket, Krabi, Phi Phi Islands), Malaysia (Penang, Langkawi), Sri Lanka (Galle), India (Goa, Kerala), and the Gulf states. Of all the states in the IOR, Myanmar has the best potential to develop eco-tourism and to conserve her relatively pristine landscape and seascape. Due to its long term political instability and its closed-door policy to foreigners and tourism, the country has an intact natural heritage of eight distinct ecosystems (Nuwer, 2016, p. 46) which could augment its village economy provided they are managed sustainably.

### *Urban development and sustainable cities*

The aforementioned developmental corridors should involve cities as well. With 55 percent of the world population living in urban areas in 2018 and a projected increase to 68 percent in 2050, urban living is the norm for human beings (UN Department of Economic and Social Affairs, Population Division, 2019, p. xix). Increasing urbanization and human population will lead to increase in demand for water, food, energy, and housing, thereby intensifying urban areas' dependence on the environment, including climate. For example, according to the FAO, the world's gross agriculture output (in terms of value) is projected to increase by around 40–53 percent between 2012 and 2050, depending on different modeling scenarios (FAO, 2018, pp. 96–98). While technological changes and innovations in agriculture have resulted in increased agriculture output in the past, at times at the expense of the environment, whether this increase can keep pace with global population growth and urban expansion remains to be seen.

Cities present formidable challenges to climate change adaptation, including its relationship to the rural agriculture sector, pre-existing brown issues, the excessive scale of mega cities, and the potential multifarious climate impacts that they face. Agriculture still remains an important economic sector for the IOR. In parts of Southeast Asia and South Asia, the productive, labor intensive agricultural sector has kept populations within rural areas and prevented the relative flood of rural migrants into cities. The productive and elastic agricultural sector has absorbed population growth in Java, creating what Clifford Geertz (1963) calls agricultural involution, by maintaining a fragile social stability. There is large underemployment but scarce unemployment and the equalization of poverty that does not lead to stark inequality as in urban areas. The term has now become synonymous with urban areas as well. With dense, congested populations, cities in the developing world now suffer from urban involution – high underemployment, low productivity, and inefficient labor-inflated bureaucracies. Urban involution is ironically a social safety valve against urban instability and revolution. This is made possible by an inflated bureaucracy, a labor absorbing service sector and the '*pasar* sector' or the 'bazaar' informal economy.

The crux of creating eco-cities and sustainable urban living lies in tackling 'brown issues' such as sewerage treatment, garbage disposal, clean water, and public hygiene, as well as implementing sound urban planning and public transport. Although the proportion of urban population living in slums have decreased from 1990 to 2014 at 28.4, 31.3, and 55.9 percent for Southeast Asia, South Asia, and Sub-Saharan Africa respectively, the urban slum population has increased (UN Habitat, 2016, p. 203). The attention to public housing can be one means of upgrading living standards, improving the quality of life, enhancing 'green living,' and spurring development. Singapore is a good example where 82 percent of population live in public housing which has enhanced living standards and maintained a quality labor force. Sustained and affordable public housing is a panacea for removing urban blight of slums and squatters (Savage & Kong, 1993). In comparison, urban greening seems like a luxury, but it should be included in the early stages of urban planning for greening to be omnipresent (Tan, Wang, & Sia, 2013).

Given the increasing trend towards megacities (cities in excess of 10 million population), the Indian Ocean states should recognize that megacities are unhealthy developments. Megacities are unsustainable agglomerations socially, economically, politically, and environmentally. Cities are bastions of civilization, and one needs to keep them as creative cultural icons, and entities for human relationships and interactions. More important, cities have to be sustainable entities without wrecking environmental destruction in surrounding ecosystems. While cities are lauded as hallmarks of human ingenuity, creativity, and innovations (Glaeser, 2011), they have been dismal failures as sustainable models of human living. Whether one adopts the Western model of 'sustainable development' (World Commission on Environment and Development, 1987) or the Chinese model of 'ecological civilization' (Savage & Lin, 2018) both are mere conceptual propositions that lack realistic successful implementation. Even for a highly planned city like Singapore (Heng, 2017), too much attention has been given to its intra-urban spatial layout, national identity, and economic viability, and little attention paid to containing its ecological footprint which makes it an unsustainable urban proposition. Nevertheless, since Singapore's growth is limited by its island geography and possible extent of land reclamation, it is an example of how cities can develop within geographic limits.

Despite all the digital paraphernalia, motor vehicles, and information and communication technology, the city should still be a humanized place, where interactions remain people to people, and ideas germinate from face to face discussions. From the first cities 10,000 years ago developed in Asia, we must not forget the objective of urban living is 'bringing us together' (Ratti & Biderman, 2017, p. 59).

Last but not least, cities face a range of climate change impacts, including sea level rise, coastal and inland flooding, drought and water scarcity, exacerbated urban heat island effects, and human health issues (IPCC, 2014b, pp. 550–563). For cities in the IOR that face severe threat of climate change impacts, resettlement, for example along developmental corridors, should be given timely and serious consideration.

## Concluding remarks

The recognition of climate change as an environmental challenge to civilizations is relatively recent in academic distillations. It was only in the last two decades that historians and social scientists have tried to apply climate change phenomena to explain the rise and fall of civilizations and expansion and contraction of kingdoms around the world (Chew, 2007; Fagan, 2008; Lieberman, 2003–2009; Lieberman & Buckley, 2012; Linden, 2006; Whyte, 2008). Given these historical correlations and interpretations, are there lessons one can learn from past experiences? The other source of pertinent sustainable experience is from contemporary indigenous societies (Diamond, 2012; Peterson, 2001; Scott, 2009).

How can we imbibe the lessons from native traditions, indigenous culture, and folk science of autochthonous societies? While we keep reciting sustainability as an objective to harmonious ecological living, the nature of modern living does not fulfill that objective. The World Wide Fund for Nature (WWF) reminds us we are spending more of nature's capital and reserves than we can afford (WWF, 2018). Scientists on the other hand alert us that 'planetary boundaries' for certain chemicals have been crossed which places Gaia in jeopardy (Foley, 2017, p. 251). Rather than wait for developed countries to take the lead in climate change mitigation and adaptation, the Indian Ocean states should step forward to deal with a regional mandate to manage its own intra-regional problems. After all, the early detection and remedial action of climate change challenges will benefit all states and communities in the IOR. The Indian Ocean states should also take pride in solving their own problems through South-South dialogue and turning to the cultural, religious, and philosophical wisdom of their societies.

## Acknowledgements

We thank two anonymous reviewers, the guest editor of this special issue Benito Cao, and the commissioning editor of the journal Adela Alfonsi for their helpful comments.

## Disclosure statement

No potential conflict of interest was reported by the authors.

## ORCID

*Lin Qi Feng* http://orcid.org/0000-0002-8442-9917

## References

Abu-Lughod, J. L. (1989). *Before European hegemony: The world system A.D. 1250–1350*. New York: Oxford University Press.

Agrawal, A. (1995). Dismantling the divide between indigenous and scientific knowledge. *Development and Change, 26*(3), 413–439. doi:10.1111/j.1467-7660.1995.tb00560.x

Andaya, B. W. (2017). Seas, oceans and cosmologies in Southeast Asia. *Journal of Southeast Asian Studies, 48*(3), 349–371.

Archer, D. (2009). *The long thaw: How humans are changing the next 100,000 years of earth's climate.* Princeton: Princeton University Press.

Bashford, A. (2017). Terraqueous histories. *The Historical Journal, 60*(2), 253–272. doi:10.1017/S0018246X16000431

Bellwood, P., Fox, J. J., & Tryon, D. (Eds.). (2006). *The Austronesians: Historical and comparative perspective.* Canberra: ANU E Press.

Berdoulay, V. (1974). *The emergence of the French School of Geography* (unpublished doctoral dissertation). Department of Geography, University of California, Berkeley.

Berkes, F. (1993). Traditional ecological knowledge in perspective. In J. T. Inglis (Ed.), *Traditional ecological knowledge: Concepts and cases* (pp. 1–9). Ottawa: International Development Research Centre.

Bicker, A., Ellen, R., & Parkes, P. (Eds.). (2003). *Indigenous enviromental knowledge and its transformations: Critical anthropological perspectives.* Florence: Taylor and Francis.

Blaut, J. M. (1993). *The colonizer's model of the world: Geographical diffusionism and eurocentric history.* New York: Guilford Press.

Briggs, J. (2005). The use of indigenous knowledge in development: Problems and challenges. *Progress in Development Studies, 5*(2), 99–114. doi:10.1191/1464993405ps105oa

Castree, N., & Gregory, D. (Eds.). (2006). *David Harvey: A critical reader.* Malden, MA: Blackwell.

Chellaney, B. (2010). Indian Ocean maritime security: Energy, environmental and climate challenges. *Journal of the Indian Ocean Region, 6*(2), 155–168. doi:10.1080/19480881.2010.536662

Chew, S. C. (2007). *The recurring dark ages: Ecological stress, climate changes, and system transformation.* Lanham: Altamira Press.

Chow, L. (2018). 2018 likely to rank as fourth-hottest year on record. *EcoWatch.*

Dallmayr, F. R. (2011). *Return to nature? An ecological counterhistory.* Lexington: University Press of Kentucky.

Diamond, J. M. (2012). *The world until yesterday: What can we learn from traditional societies?* New York: Viking.

Duara, P. (2014). *The crisis of global modernity: Asian traditions and a sustainable future.* Cambridge, UK: Cambridge University Press.

Fagan, B. M. (2008). *The great warming: Climate change and the rise and fall of civilizations.* New York: Bloomsbury Press.

FAO. (2018). *The future of food and agriculture: Alternative pathways to 2050.* Rome: FAO.

FAO, IFAD, UNICEF, WFP, & WHO. (2017). *The state of food security and nutrition in the world 2017: Building resilience for peace and food security.* Rome: FAO.

Foley, J. (2017). Living by the lessons of the planet. *Science, 356*(6335), 251–252. doi:10.1126/science.aal4863

Foster, J. B., Clark, B., & York, R. (2010). *The ecological rift: Capitalism's war on the earth.* New York: NYU Press.

Frantzi, S., Carter, N. T., & Lovett, J. C. (2009). Exploring discourses on international environmental regime effectiveness with Q methodology: A case study of the Mediterranean action plan. *Journal of Environmental Management, 90*(1), 177–186. doi:10.1016/j.jenvman.2007.08.013

Friess, D. A., Rogers, K., Lovelock, C. E., Krauss, K. W., Hamilton, S. E., Lee, S. Y., … Shi, S. (2019). The state of the world's mangrove forests: Past, present, and future. *Annual Review of Environment and Resources, 44*(1), 89–115. doi:10.1146/annurev-environ-101718-033302

Geertz, C. (1963). *Agricultural involution: The process of ecological change in Indonesia.* Berkeley: University of California Press.

Ghosh, A. (2016). *The great derangement: Climate change and the unthinkable.* Chicago: University of Chicago Press.

Glaeser, E. L. (2011). *Triumph of the city: How our greatest invention makes us richer, smarter, greener, healthier, and happier.* New York: Penguin Press.

Glover, D., & Jessup, T. (Eds.). (1999). *Indonesia's fires and haze: The cost of catastrophe*. Singapore: Institute of Southeast Asian Studies.

Green, N. (2018). The waves of heterotopia: Toward a vernacular intellectual history of the Indian Ocean. *The American Historical Review, 123*(3), 846–874. doi:10.1093/ahr/123.3.846

Han, W., Meehl, G. A., Rajagopalan, B., Fasullo, J. T., Hu, A., Lin, J., ... Yeager, S. (2010). Patterns of Indian Ocean sea-level change in a warming climate. *Nature Geoscience, 3*, 546–550. doi:10.1038/ngeo901

Haw, S. G. (2017). The maritime routes between China and the Indian Ocean during the second to ninth centuries *c.e. Journal of the Royal Asiatic Society of Great Britain & Ireland, 27*(1), 53–81. doi:10.1017/S1356186316000341

Hawksley, H. (2018). *Asian waters: The struggle over the Asia-Pacific and the strategy of Chinese expansion*. London: Duckworth Overlook.

Heng, C. K. (2017). *50 years of urban planning in Singapore*. Singapore: World Scientific.

Holland, A. (2016). Preventing tomorrow's climate wars. *Scientific American, 314*(6), 60–65. doi:10.1038/scientificamerican0616-60

Huq, N., Hugé, J., Boon, E., & Gain, A. K. (2015). Climate change impacts in agricultural communities in rural areas of coastal Bangladesh: A tale of many stories. *Sustainability, 7*(7), 8437–8460. Retrieved from https://www.mdpi.com/2071-1050/7/7/8437

Ibrahim, A. (2010). *The Asian renaissance*. Singapore: Marshall Cavendish International.

Imanishi, K. (2002). *A Japanese view of nature: The world of living things* (P. J. Asquith, H. Kawakatsu, S. Yagi, & H. Takasaki, Trans.). New York, NY: RoutledgeCurzon.

IPCC. (2014a). *Climate change 2013: The physical science basis; working group I contribution to the fifth assessment report of the intergovernmental panel on climate change*. Cambridge: Cambridge University Press.

IPCC. (2014b). *Climate change 2014: Impacts, adaptation, and vulnerability; part A: Global and sectoral aspects; contribution of working group II to the IPCC fifth assessment report*. Cambridge: Cambridge University Press.

IPCC. (2014c). *Climate change 2014: Impacts, adaptation, and vulnerability; part B: Regional aspects; contribution of working group II to the IPCC fifth assessment report*. Cambridge: Cambridge University Press.

IPCC. (2019). Summary for policymakers. In H.-O. Pörtner, D. C. Roberts, V. Masson-Delmotte, P. Zhai, M. Tignor, E. Poloczanska, K. Mintenbeck, M. Nicolai, A. Okem, J. Petzold, B. Rama, & N. Weyer (Eds.), *IPCC special report on the ocean and cryosphere in a changing climate*. In press.

IPCC. (2018). Summary for policymakers. In V. Masson-Delmotte, P. Zhai, H.-O. Pörtner, D. Roberts, J. Skea, P. R. Shukla, A. Pirani, W. Moufouma-Okia, C. Péan, R. Pidcock, S. Connors, J. B. R. Matthews, Y. Chen, X. Zhou, M. I. Gomis, E. Lonnoy, T. Maycock, M. Tignor, & T. Waterfield (Eds.), *Global warming of 1.5°C. An IPCC special report on the impacts of global warming of 1.5°C above pre-industrial levels and related global greenhouse gas emission pathways, in the context of strengthening the global response to the threat of climate change, sustainable development, and efforts to eradicate poverty*. Retrieved from https://www.ipcc.ch/sr15/

Kirch, P. V. (2010). Controlled comparison and Polynesian cultural evolution. In J. M. Diamond & J. A. Robinson (Eds.), *Natural experiments of history* (pp. 15–52). Cambridge, MA: Belknap Press of Harvard University Press.

Kirk, W. (1951). Indian Ocean community. *Scottish Geographical Magazine, 67*(3-4), 161–177. doi:10.1080/00369225108735490

Kolbert, E. (2014). *The sixth extinction: An unnatural history*. New York: Henry Holt and Company.

Lansing, J. S. (1991). *Priests and programmers: Technologies of power in the engineered landscape of Bali*. Princeton, NJ: Princeton University Press.

Lee, S.-K., Park, W., Baringer, M. O., Gordon, A. L., Huber, B., & Liu, Y. (2015). Pacific origin of the abrupt increase in Indian Ocean heat content during the warming hiatus. *Nature Geoscience, 8*, 445–449. doi:10.1038/ngeo2438

Lieberman, V. (2003–2009). *Strange parallels: Southeast Asia in global context, c. 800–1830*. New York: Cambridge University Press.

Lieberman, V., & Buckley, B. (2012). The impact of climate on Southeast Asia, circa 950–1820: New findings. *Modern Asian Studies, 46*(5), 1049–1096. doi:10.1017/S0026749X12000091

Linden, E. (2006). *The winds of change: Climate, weather, and the destruction of civilizations*. New York: Simon & Schuster.

Linton, R. (1955). *The tree of culture*. New York: Alfred A. Knopf.

Llewellyn, L. E., English, S., & Barnwell, S. (2016). A roadmap to a sustainable Indian Ocean blue economy. *Journal of the Indian Ocean Region, 12*(1), 52–66. doi:10.1080/19480881.2016.1138713

Luo, J.-J., Sasaki, W., & Masumoto, Y. (2012). Indian Ocean warming modulates Pacific climate change. *Proceedings of the National Academy of Sciences, 109*(46), 18701–18706. doi:10.1073/pnas.1210239109

Mahathir Warns Against New 'Colonialism' During Visit to China. (2018). Retrieved from https://www.bloomberg.com/news/articles/2018-08-20/mahathir-warns-against-new-colonialism-during-visit-to-china

Malinowski, B. (1922). *Argonauts of the Western Pacific: An account of native enterprise and adventure in the archipelagoes of Melanesian New Guinea*. London: George Routledge and Sons.

McGranahan, G., Balk, D., & Anderson, B. (2007). The rising tide: Assessing the risks of climate change and human settlements in low elevation coastal zones. *Environment and Urbanization, 19*(1), 17–37.

McPhaden, M. J., Meyers, G., Ando, K., Masumoto, Y., Murty, V. S. N., Ravichandran, M., … Yu, W. (2009). RAMA: The research moored array for African–Asian–Australian monsoon analysis and prediction*. *Bulletin of the American Meteorological Society, 90*(4), 459–480. doi:10.1175/2008bams2608.1

Mercer, J., Dominey-Howes, D., Kelman, I., & Lloyd, K. (2007). The potential for combining indigenous and western knowledge in reducing vulnerability to environmental hazards in small island developing states. *Environmental Hazards, 7*(4), 245–256. doi:10.1016/j.envhaz.2006.11.001

Michel, D. (2012). Environmental pressures in the Indian Ocean. In D. Michel & R. Sticklor (Eds.), *Indian Ocean rising: Maritime security and policy challenges* (pp. 113–129). Washington, DC: Stimson.

Mishra, P. (2012). *From the ruins of empire: The revolt against the west and the remaking of Asia*. London; New York: Allen Lane: an imprint of Penguin Books.

Mochizuki, T., Kimoto, M., Watanabe, M., Chikamoto, Y., & Ishii, M. (2016). Interbasin effects of the Indian Ocean on Pacific decadal climate change. *Geophysical Research Letters, 43*(13), 7168–7175. doi:10.1002/2016GL069940

Nuwer, R. (2016). Saving Eden. *Scientific American, 314*(5), 44–49.

Oka, I. N. (1991). Success and challenges of the Indonesia national integrated pest management program in the rice-based cropping system. *Crop Protection, 10*(3), 163–165. doi:10.1016/0261-2194(91)90037-R

Okumu, W. A. J. (2002). *The African renaissance: History, significance, and strategy*. Trenton, NJ: Africa World Press.

Pascoe, B. (2018). *Dark Emu: Aboriginal Australia and the birth of agriculture*. Melbourne: Scribe.

Passmore, J. A. (1980). *Man's responsibility for nature: Ecological problems and western traditions* (2nd ed.). London: Duckworth.

Peterson, A. L. (2001). *Being human: Ethics, environment, and our place in the world*. Berkeley: University of California Press.

Piketty, T. (2014). *Capital in the twenty-first century* (A. Goldhammer, Trans.). Cambridge, MA: Belknap Press of Harvard University Press.

Pimentel, D. (1996). Green revolution agriculture and chemical hazards. *Science of The Total Environment, 188*, S86–S98. doi:10.1016/0048-9697(96)05280-1

Pollock, S. I. (2006). *The language of the gods in the world of men: Sanskrit, culture, and power in premodern India*. Berkeley: University of California Press.

Pope Francis. (2015). *Laudato Si': On care for our common home*. Retrieved from http://w2.vatican.va/content/francesco/en/encyclicals/documents/papa-francesco_20150524_enciclica-laudato-si.html

Ptolemy, C. (1932). *Geography of Claudius Ptolemy* (E. L. Stevenson, Trans.). New York: New York Library.

Ratti, C., & Biderman, A. (2017). From parking lot to paradise. *Scientific American, 317*(1), 54–59.

Ray, H. P., & Alpers, E. A. (Eds.). (2007). *Cross currents and community networks: The history of the Indian Ocean world*. New Delhi; New York: Oxford University Press.

Rees, M. J. (2003). *Our final century: Will the human race survive the twenty-first century?* New York: Basic Books.

Roxy, M. K., Ritika, K., Terray, P., & Masson, S. (2014). The curious case of Indian Ocean warming. *Journal of Climate, 27*(22), 8501–8509. doi:10.1175/jcli-d-14-00471.1

Ruane, A. C., Major, D. C., Yu, W. H., Alam, M., Hussain, S. G., Khan, A. S., ... Rosenzweig, C. (2013). Multi-factor impact analysis of agricultural production in Bangladesh with climate change. *Global Environmental Change, 23*(1), 338–350. doi:10.1016/j.gloenvcha.2012.09.001

Said, E. W. (1978). *Orientalism*. New York: Pantheon Books.

Said, E. W. (1993). *Culture and imperialism*. New York: Knopf.

Savage, V. R. (1997). Singapore's garden city: Translating environmental possibilism. In G. L. Ooi & K. Kwok (Eds.), *City and the state: Singapore's built environment revisited* (pp. 187–202). Singapore: Institute of Policy Studies, Oxford University Press.

Savage, V. R. (2004). Singapore's environmental ideology. In K. C. Ban, A. Pakir, & C. K. Tong (Eds.), *Imagining Singapore* (pp. 210–239). Singapore: Eastern Universities Press.

Savage, V. R., & Kong, L. (1993). Urban constraints, political imperatives: Environmental 'design' in Singapore. *Landscape and Urban Planning, 25*(1), 37–52. doi:10.1016/0169-2046(93)90121-S

Savage, V. R., & Lin, Q. F. (2018). 'Ecological civilisation': China's ideological counter-view? *RSIS Commentary, CO18122*. Retrieved from https://www.rsis.edu.sg/rsis-publication/rsis/co18122-ecological-civilisation-chinas-ideological-counter-view/

Scott, J. C. (1976). *The moral economy of the peasant: Rebellion and subsistence in Southeast Asia*. New Haven: Yale University Press.

Scott, J. C. (2009). *The art of not being governed: An anarchist history of upland Southeast Asia*. New Haven: Yale University Press.

Sembiring, M. (2018). *The case for a dedicated regional mechanism for climate change: A comparative assessment*. Singapore: RSIS Centre for Non-Traditional Security Studies (NTS Centre).

Singh, A. (2018). India needs a better plan in the Indian Ocean. *East Asian Forum*. Retrieved from http://www.eastasiaforum.org/2018/05/12/india-needs-a-better-plan-in-the-indian-ocean/

Sivaram, V. (2017). The global warming wild card. *Scientific American, 316*(5), 48–53.

Skjærseth, J. B. (2002). The effectiveness of the Mediterranean action plan. In *Environmental regime effectiveness: Confronting theory with evidence* (pp. 311–330). Cambridge, MA: MIT Press.

Sollewijn Gelpke, J. H. F. (1995). Afonso de Albuquerque's pre-Portuguese 'javanese' map, partially reconstructed from Francisco Rodrigues' book. *Bijdragen tot de Taal-, Land- en Volkenkunde / Journal of the Humanities and Social Sciences of Southeast Asia, 151*(1), 76–99.

Sopher, D. E. (1977). *The sea nomads: A study of the maritime boat people of Southeast Asia*. Singapore: National Museum.

Tal, A. (2007). To make a desert bloom: The Israeli agricultural adventure and the quest for sustainability. *Agricultural History, 81*(2), 228–257. Retrieved from http://www.jstor.org.ezlibproxy1.ntu.edu.sg/stable/4617826

Tan, P. Y., Wang, J., & Sia, A. (2013). Perspectives on five decades of the urban greening of Singapore. *Cities, 32*, 24–32.

Tortajada, C., Joshi, Y., & Biswas, A. K. (2013). *The Singapore water story: Sustainable development in an urban city-state*. London: Routledge.

UN Department of Economic and Social Affairs, Population Division. (2019). *World urbanization prospects: The 2018 revision (ST/ESA/SER.A/420)*. New York: United Nations.

UN Habitat. (2016). *Urbanization and development: Emerging futures; world cities report 2016*. Nairobi: UN Habitat.

Warmington, E. H. (1974). *The commerce between the Roman Empire and India* (2nd ed.). London: Curzon.

Wei, T., Dong, W., Yan, Q., Chou, J., Yang, Z., & Tian, D. (2016). Developed and developing world contributions to climate system change based on carbon dioxide, methane and nitrous oxide emissions. *Advances in Atmospheric Sciences, 33*(5), 632–643. doi:10.1007/s00376-015-5141-4

White, L., Jr. (1967). The historical roots of our ecologic crisis. *Science, 155*(3767), 1203–1207. Retrieved from http://www.jstor.org/stable/1720120

Whyte, I. (2008). *World without end: Environmental disaster and the collapse of empires*. London: I.B. Tauris.
World Commission on Environment and Development. (1987). *Our common future*. Oxford: Oxford University Press.
WWF. (2018). *Living planet report – 2018: Aiming higher*. Gland, Switzerland: WWF.
Xie, J., Yu, J., Chen, B., Feng, Z., Lyu, J., Hu, L., … Siddique, K. H. M. (2018). Gobi agriculture: An innovative farming system that increases energy and water use efficiencies. A review. *Agronomy for Sustainable Development, 38*(6), 62. doi:10.1007/s13593-018-0540-4

# Indian Ocean fisheries regulation: exploring participatory approaches to support small-scale fisheries in six States

Erika Techera

**ABSTRACT**

In the Indian Ocean region fish and other marine living resources are vital for food security, local livelihoods, national economies, and future development opportunities. Of particular importance is the small-scale fishing sector, which supplies food and employment for many local people and coastal communities. To ensure the longevity of artisanal and subsistence fisheries, and healthy oceans and marine living resources, improved legal governance is critical. Most Indian Ocean States have adopted fisheries laws to achieve this goal, yet successfully balancing conservation and sustainable use of marine resources remains a challenge. To enhance fisheries laws in efficient, effective and equitable ways, inclusive governance approaches are needed. This article analyses fisheries laws in six Indian Ocean States, and critically assesses the incorporation of participatory approaches that support small-scale fisheries. The comparative analysis provides a resource that will assist States in identifying legal options to enhance public participation and inclusive fisheries governance.

## Introduction

The Indian Ocean region includes both small island States and large continental countries, as well as vast ocean areas. There is no settled list of countries, but the most inclusive regional organization, the Indian Ocean Rim Association (IORA), comprises 21 independent nations spanning Africa and Asia (IORA, 2019).[1] The region also includes remarkable economic and cultural diversities; comprising least developed, developing countries, and developed nations, and almost every legal system in the world from common and civil law, to customary and sharia laws (Techera, 2019).

A key feature of the Indian Ocean region is the socio-economic dependence placed upon the ocean. Fish and other marine living resources have been a vital source of food for many coastal communities for millennia, and the fishing sector continues to provide food and valuable livelihoods, as well as contributing significantly to economies and development opportunities (Mohanty, Dash, Gupta, & Gaur, 2015; van der Elst et al., 2005; Voyer et al., 2018). Of particular importance is the small-scale fisheries sector, which supplies employment for large numbers of local people in countries such as

Tanzania and Sri Lanka (Jiddawi & Öhman, 2002; Pathmanandakumar, 2017). In other States, such as Seychelles, the sector is smaller, but the average consumption of fish is one of the highest in the world (Clifton et al., 2012). In the last decade, many Indian Ocean States have adopted blue economy policies, which in part involve goals focused on expanding fisheries sectors; both wild-caught and farmed fish (Doyle, 2018; Govt of Seychelles, 2017; IORA, 2017; and Nairobi Statement, 2018). Blue economy goals seek to harness the benefits of increased food production, employment and economic development opportunities (Mohanty et al., 2015); yet it is clear that indicators for many wild fish stocks are worsening and harvests are at or close to their maximum sustainable yield (FAO, 2018a). In addition, fish stocks will be impacted by the changing marine environment: directly as oceans warm and/or acidify, or indirectly through the impacts of extreme weather events and coastal erosion, leading to loss of breeding and nursery grounds as well as coastal infrastructure (Bijma, Pörtner, Yesson, & Rogers, 2013; Blasiak et al., 2017; FAO, 2018b; Lan, Evans, & Lee, 2013; Singh, 2015; and Techera, 2018).

Small-scale fisheries play a critical role in providing food, livelihoods and economic security, including in the Indian Ocean region (Cohen et al., 2019; FAO, 2018a). Increasing pressure from fishing, environmental degradation, and climate change, plus planned expansion and blue economy goals, bring sharply into focus the need for enhanced fisheries governance. Advancing more participatory governance approaches is one way forward. Public participation in governance is widely regarded as beneficial and is considered uncontroversial (Coffey, 2005, p. 31). This research focuses on the three elements of public participation in decision-making, access to information and access to justice, as they are well endorsed at the international and national levels, as set out below. Each of these elements is inter-connected. For example, whilst public participation can be facilitated through membership of governance bodies, various decision-making processes, and in ongoing fisheries management, in order to participate fully fishers – and the broader public – require access to the information upon which decisions are made, rights of standing before the courts and the authority to challenge processes and decisions.

It is in this context that this article examines the extent to which the national fisheries laws in six Indian Ocean States support participation in fisheries governance. The purpose is not to assess the success of the legal frameworks on the ground, but rather to identify whether and to what extent small-scale fishers have opportunities to participate in decision-making, and have access to information and justice. The paper commences by outlining the rationale for the research and the methodology adopted. Thereafter, the broader context of public participation in good governance is explored, and its specific relevance to fisheries sectors, before the case studies are examined. Thereafter, the paper comparatively analyses the national fisheries laws to draw out best practice legal options to facilitate participation in fisheries governance. This research contributes to the literature in two ways: by highlighting Indian Ocean fisheries legislation in lesser-known jurisdictions and by illuminating the ways in which public participation in small-scale fisheries governance is facilitated in this region.

## Research focus and methodology

This paper provides a desk-based, doctrinal analysis of fisheries law across six Indian Ocean States, to identify the extent to which participatory approaches have been embraced to support inclusive small-scale fisheries governance. For the purposes of this paper, small-scale fishing is considered to include artisanal, traditional and subsistence fisheries, as opposed to large-scale fisheries (Pathmanandakumar, 2017). As noted above, three key pillars of participatory governance are examined: public participation itself, access to information and access to justice (see *Rio Declaration* and *Aarhus Convention*, examined below).

The analysis explores the text of national fisheries legislation in three island nations (Mauritius, Seychelles and Sri Lanka) and three littoral countries (Bangladesh, Kenya, and Tanzania). It is acknowledged that the laws as written may not align with their implementation on the ground. Furthermore, it is clear that fisheries regulations exist at multiple levels, from national legislation to local level customary rules and by-laws (Sobo, 2012), and also in legally pluralist contexts (Wickramasinghe & Bavinck, 2015). Nevertheless, examining national legal frameworks is important and valuable, as these laws have a top-down influence on other levels of governance, and can create a supportive environment for local-level participation that meets socio-cultural needs (Kearney, Berkes, Charles, Pinkerton, & Wiber, 2007). National governments are also uniquely placed to develop laws and policies that embed international standards and guidelines from global organizations such as the Food and Agriculture Organization (FAO) (e.g. FAO, 2015). Finally, analysis of the domestic legal frameworks can inform future instruments to address gaps, recommendations for reform and model laws.

Each of the six nations has large coastal areas and growing populations, with varying levels of dependency upon small-scale fisheries. These case study countries differ from each other in terms of geography, development status, legal history and location in African and Asian regions, enriching the research landscape. The jurisdictions also provide a range of different approaches to fisheries governance, with laws that vary in terms of their opportunities for participation. Therefore, their comparative analysis serves to highlight a variety of options, and legal environments, that enable public participation.

In the Indian Ocean region, relatively little research attention has been paid to fisheries laws in these jurisdictions. Therefore, this paper provides a foundational study only, and it is clear that further research is needed to expand the analysis to all nations in the region. Additional studies will also be needed to explore the operational effect of the laws, particularly as there is evidence of weak governance and non-compliance with the written law in some circumstances (Jiddawi & Öhman, 2002; Pathmanandakumar, 2017). Further research is also needed in the individual jurisdictions, because each has a unique legal and cultural environment. Nevertheless, legal literature in the field of global environmental law demonstrates that similar environmental laws have emerged from multiple jurisdictions even where they have diverse legal systems and cultures (Yang & Percival, 2009). Therefore, this comparative study serves to highlight approaches taken in a small range of States, and is a first step towards better understanding of participatory fisheries governance in the Indian Ocean.

The comparative analysis of the six case study jurisdictions focuses on five key aspects of the legislation. First, the objects and purposes of the legislation itself. This is important in terms of understanding the goals of the laws, and also the extent to which socio-cultural aspects were a focus of the instrument at the time of its adoption. Secondly, the institutions with decision-making authority, including their composition and functions. These institutions can be more or less inclusive, by permitting or requiring membership by one or many non-governmental members. Thirdly, is the issue of the extent to which the legislation supports the small-scale fishing sector, including traditional marine tenure, customary fishing rights, access to traditional fishing grounds and commitments to local fishers and fisheries more broadly. It is clear that these are all conditions for successful inclusive governance (Armitage et al., 2009). The purpose here is to provide a high-level examination of how the legislation accounts for these elements, rather than assess the specific jurisdictional nuances. Fourthly, the research identifies opportunities for access to information, consultation and public participation in fishery management and planning. Lastly, rights of standing to appeal decisions and/or challenge the law, and rights of access to justice are examined.

Before turning to the national legal frameworks, and the comparative analysis, the broader legal context of public participation in fisheries governance is considered.

## Public participation, good governance, and law

There are several reasons to enhance public participation in governance. First, good governance is often articulated in terms of its principles (UNDP, 2011). Public participation is one of these principles (UNDP, 2011), and it has been said that it is 'an essential element of good governance' (Coffey, 2005, p. 27). Governance has no globally agreed definition, but for the purposes of this paper, it is defined as incorporating the binding rules, the institutions that make and implement them, and the relevant processes involved (UNECOSOC, 2006). Therefore, legal institutions and mechanisms that incorporate participatory approaches and principles are more likely to contribute positively to good governance.

Secondly, there is widespread support for public participation in environmental development and governance of natural resources (Coffey, 2005; Hassan, 1999). In particular, to achieve blue economy goals that take into account the small-scale fisheries sector, collaborative and inclusive governance is required (Cohen et al., 2019); with appropriate legal frameworks to support participatory fisheries management and encourage its success (Haambiya, Kaunda, Likongwe, Kambewa, & Muyangali, 2015). In addition, Indian Ocean States have a long history of fishing, and there are frequently customary laws, practices, and beliefs that form an important part of the culture of coastal communities (Snijman, 2011). Consequently, inclusive approaches that recognize the rights of local fishers and traditional governance approaches are important for both good governance and social justice.

Further support for public participation in good governance can be found in international law. At the global level, States have been encouraged to adopt approaches that facilitate the broadest public participation (UNCED, 1992). Significantly, at the United Nations Conference on Environment and Development in 1992, the international community endorsed Agenda 21, which encouraged:

> Governments ... [to] establish innovative procedures, programmes, projects and services that facilitate and encourage the active participation of those affected in the decision-making and implementation process, especially of groups that have, hitherto, often been excluded, such as women, youth, indigenous people and their communities and other local communities. (UNCED, 1992, para 10.10)

This includes both decision-making and law-making, as well as specific areas such as environmental impact assessment procedures (UNCED, 1992, para 15.5), risk assessment and management (UNCED, 1992, para 16.30), awareness raising generally, and active community involvement in environmental initiatives (e.g. UNCED, 1992, para 18.59(d)). The *Rio Declaration* also notes that 'environmental decisions are best handled with the participation of all concerned citizens' and encourages States to 'facilitate and encourage public awareness and participation by making information widely available, [including] [e]ffective access to judicial and administrative proceedings' (Principle 10). Principle 22 further encourages States to enable the effective participation of 'Indigenous people and their communities and other local communities.'

The later United Nations (UN) Agenda 2030 document again emphasizes the need for participation of all people in sustainable development, including women and girls (UNGA, 2015).[2] The Sustainable Development Goals (SDGs) include references to participation several times, most relevantly in SDG16, which focuses on efforts to '[p]romote peaceful and inclusive societies for sustainable development, provide access to justice for all and build effective, accountable and inclusive institutions at all levels' including through ensuring 'responsive, inclusive, participatory and representative decision making at all levels.' In addition to these soft law developments, the *Aarhus Convention on Access to Information, Public Participation in Decision-Making and Access to Justice in Environmental Matters* provides binding rules in relation to public participation in decision-making as well as access to information and justice. Although this is a regional convention binding only on European member States, it articulates Principle 10 of the Rio Declaration and is described as having 'global' scope (Okello, Beevers, Douven, & Leentvaar, 2009).

Inclusive governance at the domestic level involves identifying opportunities and facilitating civil society engagement and public participation in institutions, laws, and processes. The various types of participation, however, are not all the same. Public participation can be passive (weak) where only basic information is provided by decision-makers and/or simple consultation required, or active (strong) when stakeholders are fully involved in functional decision-making, empowered to take a partnership role in management or where authority is fully devolved. It is clear that there is a considerable 'difference between going through the empty ritual of participation and having the real power needed to affect the outcome of the process' (Arnstein, 1969, p. 216). Frequently, legal frameworks include a range of provisions across this spectrum. For example, legislation may provide for weak participation where the public is merely informed and/or consulted on various issues affecting them. In other contexts, higher-level participation is supported, involving collaboration, partnership and sometimes complete devolution of authority to make decisions. Co-management and community-based management, therefore, sit at the upper end of this spectrum as they involve government-community partnerships and/or complete control of management decisions about natural resources or areas being given to local groups (Pathmanandakumar, 2017).

The operationalization of participatory governance is, however, more complex than the above summary suggests, involving engagement with multiple actors, operating at various levels (Bavinck, Jentoft, Pascual-Fernández, & Marciniak, 2015). Legal frameworks can play a critical role by providing a foundation that supports or hinders the multi-faceted nature of good governance. Laws can embed good governance principles to enable their achievement and to provide safeguards for stakeholders. Furthermore, it has been noted that blue economy goals, and broader development objects, can only be achieved if effective legal regimes are in place (Llewellyn et al, 2016; and Mohanty et al., 2015). Therefore, there are several reasons to focus on enhancing legislation to achieve blue economy outcomes and participatory fisheries governance.

In the context of the fisheries sector, most Indian Ocean States have fisheries legislation that seeks to sustainably conserve and manage fish stocks and fishing activities (Snijman, 2011). Increasingly, however, scientific evidence is demonstrating that fish stocks are threatened by over-harvesting and other environmental factors, which necessitate improved, more effective, governance and therefore laws. At its simplest, to be effective, laws must achieve their objects (McGrath, 2010). There are, however, multiple ways in which goals can be realized, including through mechanisms that are more participatory and involve multiple stakeholders. Some commentators believe that State intervention can be superfluous to local governance in some circumstances (Wickramasinghe & Bavinck, 2015). Other research has demonstrated that governance is too complex to be governed by a single agency (Berkes, 2008), and that small-scale fisheries will be unsuccessful if controlled by a single institution (Pathmanandakumar, 2017). Yet in other situations, 'mixed arenas' can be challenging for local fishers and fishing communities (Delgado-Serrano, Mistry, Matzdorf, & Leclerc, 2017). Inclusive approaches to fisheries governance can address these issues, in part, by involving multiple actors at various stages of decision-making processes, and in different ways.

Effectiveness also requires laws to be implemented, complied with and enforced. Non-compliance and lack of enforcement are common challenges in fisheries governance, including in this region (Sobo, 2012). To support successful implementation of fisheries laws, opportunities must be found to involve local people in governance (including fishers and the broader public), to build environmental citizenship and natural resource stewardship values. Encouraging compliance, and discouraging non-compliance, is critical to improving governance, because the regulatory costs of monitoring and enforcement are high (Stohr, Lundholm, Crona, & Chabay, 2014). Compliance tends to be reduced in situations of conflict. Public participation can assist in resolving conflicts and accommodating differing views (Coffey, 2005). Public participation can also increase compliance if it is well-structured and transparent, participants are involved in genuine decision-making, and there is space for compromise (Coffey, 2005, p. 31). This again points to the necessity for inclusive governance approaches through public participation in institutions, at all stages of law-making and implementation, as well as enforcement.

## Embedding public participation in fisheries governance

There has been some literature that addresses public participation in fisheries governance (e.g. Coffey, 2005; Stohr et al., 2014; Suárez de Vivero, Rodríguez Mateos, & Florido del Corral, 2008; and Wiser, 2000) and some that specifically focuses on small-scale fisheries in the region (e.g. Pathmanandakumar, 2017; Wickramasinghe & Bavinck, 2015). In addition, the international community has adopted an international framework for action, and targets, that include sustainable fisheries. Relevantly, SDG14.B makes specific reference to providing access to resources and markets for small-scale artisanal fishers. The Food and Agriculture Organization (FAO) of the United Nations recognizes the importance of small-scale fisheries and has committed to supporting SDG14 especially with respect to awareness raising, strengthening institutions, empowering fishery organizations, supporting policy reform and providing technical assistance to implement FAO's *Voluntary*

*Guidelines for Securing Sustainable Small-Scale Fisheries in the Context of Food Security and Poverty Eradication* (FAO, 2018a, p. 88). These Guidelines include consultation and 'active, free, effective, meaningful and informed' participation as guiding principles (FAO, 2015, p. 3). Specific reference is also made to participation in sustainable resource management, social development, and work, as well as securing the specific participation of women. Relevantly, the Guidelines encourage States to take into account small-scale fisheries interests in designing laws and policies, through consultation, participation, and publication.

Overall, it is clear that there are considerable benefits to having broad opportunities for public participation in fisheries governance. Top-down, centralized governance, by itself, is unlikely to be effective given the high regulatory costs of monitoring and enforcement. Nevertheless, bottom-up approaches are not a panacea. For example, focusing only on localized community-based fisheries management may not achieve national policy goals, or be able to address complex legal issues surrounding issues such as property rights (Pathmanandakumar, 2017; Stohr et al., 2014). The answers to these issues lie in combining top-down and bottom-up approaches, and integrating local and national governance, in a collaborative framework that fully engages with fishers. Getting the balance right is no easy task for law-makers who must draft and implement fisheries laws. If legal frameworks impinge on local fishers' rights and expectations, tensions can arise to which governments must respond. Simultaneously, regulators must address broader civil society interests about the environmental and natural resource impacts of fishing. In order for governments to understand, engage with and respond to fishing and non-fishing community concerns, ways must be found for civil society participation in institutions, decision-making processes, and sustainable fisheries management. For this to be achieved in efficient, effective and equitable ways, fishery laws must incorporate best practice mechanisms for public participation.

It is evident that public participation can be facilitated in legislation in several ways. For example, at the institutional level, representative members of the fishing sector and/or broader public can be involved in fisheries advisory and management bodies. Processes can be put in place that require access to information and community consultation, and that permit submissions to be made, and considered, on law and management of fisheries. Management can be centralized, collaborative or fully devolved to local fishing communities. Legislation can also create certainty around access, ownership and management rights, as well as rights to information, participation, and justice.

The following analysis focuses on the specific context of six Indian Ocean States. It is acknowledged that in many respects these jurisdictions differ considerably, in terms of their history, natural environments, legal systems, cultural contexts, and economic development objectives. Nevertheless, as discussed above, there is much to be learned from a high-level exploration of the ways in which participation is embedded, or not, in fisheries legislation. The sections that follow explore each case study country in turn.

## National fisheries legislation

### *Bangladesh*

Bangladesh is an example of a State with older fisheries legislation that was developed before the participatory principles and approaches to governance, outlined above, were widely endorsed. There are two relevant pieces of legislation in Bangladesh. The first is the *Protection and Conservation of Fish Act 1950*. This Act allows the government to regulate certain fishing and fishery-related activities, but does not provide for fishery management and planning. It includes no reference to small-scale, traditional or subsistence fishing rights, no opportunities for public participation in institutions or management planning, and provides no access to information or justice. This is not surprising given the very early date of the statute.

The *Marine Fisheries Ordinance 1983* is the more relevant instrument, focused on 'provisions for the management, conservation, and development of marine fisheries of Bangladesh.' It has no objects clause or further explanation to evidence strong support for local fishers or livelihoods. Nevertheless, there are some relevant provisions, explored below.

The Ordinance does not establish any advisory or fishery management body. Responsibility is given to the Director 'for the management, conservation, supervision and development of marine fisheries and the implementation of the objectives of this Ordinance' (s.5). Furthermore, the government is empowered to specify the types and numbers of vessels that can be deployed in terms of fisheries management and development (s.7). No specific mention is made of traditional, artisanal or subsistence fishing rights. Nevertheless, some provisions do provide for local fisheries, however, the Director can impose additional conditions on a foreign fishing vessel license in order to protect local fisheries (s.16(3)(c)). In general, local fishing vessels must be licensed, but the government may grant exemptions from licensing requirements for such vessels, and declare zones where only non-mechanized or mechanized local fishing vessels are permitted to fish or other fishing restricted (s.3). These provisions could be utilized to protect local livelihoods and traditional fishing rights, but without explicit guiding principles this is not assured.

The Ordinance gives power to the government to make rules for, *inter alia*, the regulation and management of marine fisheries, catch and size limits, licensing, area closures (s.55). No mention is made of consultation and participation in the Ordinance itself or the subsequent Rules. The *Marine Fisheries Rules 1983*, exempt local non-mechanized vessels from license fees, providing some special treatment for local fishers (rule 6). A further limited effort is made to secure livelihoods, as the Rules provide that at least three Bangladeshi citizens must be employed on each licensed trawl fishing vessel (rule 18). These provisions provide no opportunities for public participation in decision-making. This may be partly accounted for as local-level fisheries management plays an important role in this country, yet national frameworks ought to provide an inclusive foundation for lower levels of governance, as discussed above.

Under Section 25 of the Ordinance, a person aggrieved by a decision not to grant, or to suspend, a license, may appeal that decision. Although this is a valuable provision, it does not provide broad access to justice for fishers or other citizens to ensure the Ordinance is complied with.

### *Kenya*

Kenya is a jurisdiction at the other end of the spectrum to Bangladesh in terms of the age, inclusivity, and comprehensiveness of its national fisheries law. The principal piece of legislation is the *Fisheries Management and Development Act 2016*, designed, in part, 'to enhance the livelihood of communities dependent on fishing.' Importantly, the Act sets out objects, including to 'uplift the living standards of the fishing communities and to introduce fishing to traditionally non-fishing communities and to enhance food security' (s.5(1)). Eighteen broad principles underpin the statute, including 'encouraging the participation of users of the fisheries resources, and the general community, in the management of fisheries' (s.5(2)(g)), and 'ensuring that the livelihood of fishers is enhanced' (s.5(2)(r)). This provides a solid foundation for public participation in fisheries governance.

In terms of institutions, the Kenya Fisheries Advisory Council is established to advise the government on policy, allocation of resources, research, education and management plans (s.6(2)). Its composition includes representatives from five national government agencies, and one person from each of a fisheries research institute/university, a consumer group, the Council of Governors and local government, and an umbrella fisheries body (s.6 (1)). The only relevant function of the Kenya Fisheries Service, for the purposes of this paper, is to 'liaise as appropriate with agencies and persons, including stakeholders, industry, government agencies, regional and international organisations and experts' (s.9(j)), and 'facilitate investment in commercial fisheries, in collaboration with relevant agencies, persons or bodies' (s.9(p)). The Service is led by a Board which includes government officers and five members 'not being public officers, openly appointed by the Cabinet Secretary for their long-term knowledge and experience in the conservation, management, development and sustainable use of fisheries resources' (s.11(1)(c)). Although extending beyond government officials, this provision would not appear to envisage local fishers or community representatives as members must have a university degree in a relevant discipline (s.11(5)). The Board can, however, invite a person to attend or make a submission (s.14(6)), and shall

> establish a consultative process with the stakeholders in order that information and views on relevant fisheries management may be exchanged as appropriate, and the Board shall take into account any information or views received from stakeholders. (s.14(7))

Significantly, the Act recognizes traditional rights which may support stewardship values. Whilst the ownership of fisheries resources is vested in the State (s.29(1)), this does not remove customary community use rights which are preserved under s.29(2). Customary fishing rights are safeguarded from seasonal, species and method restrictions, as well as limitations on vessel sizes (ss.40(1)(a) and (d)). Traditional fishing rights and consultation are further protected under s.64(1), which provides that '[n]o person shall, by carrying out aquaculture activities, deprive a local community of its traditional access to fishing grounds without good cause and without first consulting the affected community.'

Participatory approaches are evident in the definition of management under the Act, as 'an integrated process of information gathering, analysis, planning, consultation, decision making, allocation of resources, formulation and implementation of rules and regulations' (s.2). Fishery management is largely devolved to counties who collaborate at the national level (s.33), but also prepare county-level fishery management measures and plans (s.34). The Board can, however, declare a designated fishery on cultural grounds and thereafter assume responsibility for it, albeit in cooperation with the county and having consulted with other stakeholders (s.39). This ought to permit national intervention where national cultural significance demands it. Importantly, guidelines and standards for the development of fisheries management plans for these fisheries must take account of 'relevant traditional fishing rights, methods or principles' and involve a 'social impact assessment of the plan with reference to disadvantaged groups including women, persons with disability and the youth' (s.39(3)(f) and (n)). The only other jurisdiction mentioning social impact, amongst the laws analyzed in this article, is Mauritius. Consultation features in several provisions which specify 'stakeholders' but do not set out precisely what that means (e.g. s.39 (4)). Stakeholders must be consulted on the declaration of any marine protected area (s.47 (1)), and information must be provided to the public and stakeholders on fisheries conservation, management, and development under s.83. The distinction between the public and stakeholders is interesting: information must be available to the public but positively disseminated to stakeholders. A stronger provision would require active dissemination to all citizens. Significantly, management measures in fishery management plans are legally binding (s.39(4)).

Beach management units are a key concept in Kenyan fisheries management, defined as organizations 'of fishers, fish traders, boat owners, fish processors and other beach stakeholders who traditionally depend on fisheries activities for their livelihoods' (s.2). Regulations may be made 'for purposes of ensuring structured community participation in fisheries management, make regulations setting out standards for the management of beach management units established by the county governments' (s.37(1)). Notably, regulations made in relation to beach management units can provide for the 'protection of vulnerable groups, especially youth and women' (s.37(e)) and 'processes necessary to ensure that not more than two-thirds of Beach Management Units are of the same gender and to ensure the inclusion of youth and persons with disability in leadership' (S.37(f)). These provisions strongly support local fishers' involvement in management. The *Fisheries (Beach Management Units) Regulations 2007* expand on these provisions, noting that beach management units are non-political and non-religious bodies aimed at strengthening fisheries management, supporting sustainable development, recognizing the roles played by different sections of the community including women, building capacity to effectively manage fisheries and reducing conflicts (reg.3). Significantly, the objectives include 'help alleviate poverty and improve the health, welfare, and livelihoods of the members through improved planning and resource management, good governance, democratic participation and self-reliance' (reg.3(3)(c)). In this sense, the Regulations support the underlying objects of the main Act. Co-management areas, where fisheries management activities are undertaken by beach management units and the Director, can be established following a 'consultative process' (reg.7(1)). The key tool utilized is the co-management plan which can specify closed areas, seasons, gear, and vessel licensing restrictions (reg.7(4)). Again consultation is required in the preparation of these plans. Local management responsibility is devolved to these bodies with the power to pass by-laws (reg.5(6)), including giving effect to a co-management plan (reg. 7(7)) and other protection measures (reg.8). Compliance is addressed in the Regulations through the establishment of patrol sub-committees (reg.7(8)). Leadership and supervision of the beach management unit are via an executive committee comprised of elected members distributed across boat owners, crews, traders, and others (regs.14 and 24). But the broader membership, via an assembly which is open to anyone directly dependent upon fisheries or who would be adversely affected by a fisheries closure, votes on approval of management plans and adoption of by-laws (regs.9 and 23). Furthermore, regulation 13 states that all members of a beach management unit may participate in decision-making processes. Provisions also allow for the expulsion of members of a beach management unit for behaving contrary to the objectives (reg.12). The regulations provide the more detailed rules needed to ensure diversity of membership and opportunities to participate in local fisheries governance.

Under the main Act, livelihoods are referred to, and safeguards established, in several places. For example, s.63(b) seeks to protect 'the livelihood, culture and traditions of local communities and their access to fishing grounds' by ensuring they are not affected by aquaculture development. In addition, the government can promote the development of activities by encouraging the adoption of alternative livelihoods under s.30(f). The Act goes further than some of its regional counterparts, by establishing a Fish Marketing Authority which, relevantly, must advise fisheries stakeholders on 'fish market needs and trends' and organize stakeholders 'to ensure smooth marketing of fish and fishery products' (s.200(e) and (f)). Again this ought to support fishing-related livelihoods and access to markets (a key goal under SDG14.B).

A right to appeal against a management measure that limits fishing or fishing-related activities may be made to the Cabinet Secretary (s.41(2)), and a right to appeal against a decision not to grant or renew a license is also provided (s.97). These provisions protect those directly engaged in small-scale fishing, but fall short of a broad right of standing to sue for breaches of the Act.

## *Sri Lanka*

Sri Lankan fisheries law is different again, and marked by the support provided for broad representation in institutions rather than processes. The *Fisheries and Aquatic Resources Act No.2 1996* includes provisions on the management, regulation, conservation, and development of fisheries and aquatic resources, and the implementation of international law; it does not, however, include any objects or principles. A Fisheries and Resources Advisory Council is established 'to advise the Minister on all matters relating to the management, regulation, conservation and development of fisheries and aquatic resources in Sri Lanka waters' (s.4(1)). Importantly, its composition includes over 20 government members drawn from different agencies, plus 18 other representative members (s.3(1)). The non-governmental members include representatives of fishing cooperatives and associations, at least two representatives of women engaged in fishing and six members with special knowledge and experiences related to the fisheries industry or other sciences. Although government members slightly exceed other representatives, inclusive membership is noteworthy. This should ensure that a diversity of views is presented in all decision-making processes, but they may not capture broader public opinion or non-represented stakeholder concerns.

A fishing license is required for all prescribed fishing operations in Sri Lankan waters (s.6), and local fishing boats must be registered under s.15. No specific mention is made of traditional access rights or subsistence fishing rights. One broad provision gives the Minister power to prepare a 'plan for the management, regulation, conservation, and development of fisheries and aquatic resources in Sri Lanka' (s.5); the only consultation required under that section is with the Council referred to above. Section 58 requires intra-governmental consultation, in particular requiring the central administration to ensure 'every province in charge of the subject of Fisheries is consulted on matters affecting fishing operations within such province.' The Act provides for protection measures for fish and prohibitions on the use of certain gear and equipment, but with no public involvement. Nevertheless, the fishers in any area may request that a Fisheries Management Area be designated (s.31(1)). Such Areas must have a Fisheries Management Coordinating Committee comprised of national government officers, district fisheries officers, municipal council members, and elected representatives of local fishing committees. The last of these is further described under Sections 32 and 33, where it is noted that their functions include 'carrying out social infrastructure and welfare activities with a view to improving the living standards of the fishing community' and 'engaging in such other activities as are … beneficial to the fishing community of the area' (ss.32(2)(c) and (d)). The government can appoint other members to the Fisheries Management Coordinating Committee, including representatives of fishing associations, women's groups and NGOs (s.31A). This Committee prepares the development and management plan for the Fisheries Management Area which may include division into zones, restrictions on gear and fishing methods, seasonal closures and other listed measures, but no reference is made to traditional, subsistence or local community needs and values (s.31B). The Proposals for consultation and review may form part of the plan (s.31B(2)(k)). Although the final plan must be gazetted, there appears to be no requirement to advertise the draft plan, nor to seek public input, beyond the views of those involved in the Committee itself. These provisions add to broader powers of the Minister to declare any area of Sri Lankan waters to be a fisheries reserve where 'special protection' is necessary; again there are no opportunities for public participation in selection, decision, or management of these areas (s.36). Again, participation is heavily dependent upon representation in decision-making bodies rather than inclusive processes.

The Act does not grant a broad right of standing to bring proceedings against a decision contrary to the Act, but does permit an appeal by a person whose fishing or aquaculture license is not granted or renewed (ss.12 and 43D).

### *Tanzania*

The Tanzanian *Fisheries Act 2003* is more aligned with the law in Kenya, although it is not as comprehensive. The Act does not include any specific objects or principles to guide interpretation and implementation. The Act does not establish a specific institution, but makes provision for a Director of fisheries as an advisor to the government (s.4). Relevantly, the Director is empowered to publish circulars to aid the implementation of the Act (s.4), and 'provide and disseminate information and guidance … to members of the public' (s.7), but has no obligation to do so. Yet this provision remains important, as past findings have recommended greater advertising of fisheries laws to the public (Jiddawi & Öhman, 2002, p. 526). The Director's main function is to 'promote, encourage and support all initiatives leading to the development and sustainable use of the fish stock and aquatic resources' (s.9(1)), and sets out 11 measures that may be taken. These, relevantly, include 'facilitating artisanal fishing, semi-intensive aquaculture with simple technologies and low capital investment,' and 'encouraging the involvement of stakeholders in the planning, development, and management of fishery resources.' These provisions are highly relevant as previous research has recommended the encouragement of traditional management and alternative (non-destructive) fishing practices (Jiddawi & Öhman, 2002, p. 526). The *Fisheries Regulations 2009* have built on this provision, for example, by stating that any 'investor may form [a] partnership with local seaweed farmers' (reg.45). Overall it appears that access to information and public participation is possible, but is not mandatory under the Act itself.

Traditional fishing rights are not explicitly secured under this Act. The Act does, however, try to ensure these rights are not impacted by actions taken under the legislation. One of the factors to be taken into account by the Director in adopting measures to maintain or restore fish stocks is 'the interest of local fishers such as those engaged in artisanal fisheries' (s.9(2)(d)). In addition, the provisions in relation to aquaculture development in s.11 oblige the Director to ensure 'livelihood, culture and traditions of local communities and their access to fishing ground are not affected,' and that the 'local community has access to fishing grounds.' Indeed, depriving a local community from access to such fishing grounds is an offence (s.11(2)). Furthermore, it is noted that if the Director makes a determination affecting a person's rights, s/he must give reasons for doing so (s.4(4)), and an aggrieved person has 30 days to appeal to the Minister (s.4(5)). The Minister is empowered to make regulations for artisanal fisheries, but not traditional fishing rights as such (s.57(2(dd))).

The provisions relating to management measures and plans do not require broad consultation and public participation, although some safeguards are established. All fishing requires a license, with the exception of fishing for prawns with cloth, using a rod and line or hand line from the beach, and using small cast nets (s.22); this would appear to exempt some traditional, artisanal and subsistence fishing methods. Furthermore, the Minister, in imposing fisheries management conditions, must give regard to 'ensuring that traditional practices ... and interests of indigenous people and local fishing communities which are highly dependent of fisheries resources for their livelihood' (s.17(p)). This section of the Act goes further, requiring the Minister to impose conditions 'facilitating the formation of community management units and authorized associations for the purpose of protecting and conserving fishery resources' (s.17(s)). In essence, these obligations require the consideration of, and provision for, local communities, but do not mandate public participation in management planning. As with the Kenyan legislation, the Tanzanian fisheries law permits the Director to enter into management agreements with beach management units (s.18). Further details are set out in the *Fisheries Regulations 2009*, which states that 'every fishing community shall, in collaboration with relevant village government, form Beach Management Units to provide for Collaborative Fisheries Management' (reg.25(1)), and thereafter the Director can enter into management agreements with them (reg.84). The jurisdictional area of each Beach Management Unit must be agreed upon by fishers, community organizations, village councils and local authorities (reg.133). The inclusion of village/local authorities is relevant as it overcomes, in part, the concerns expressed above regarding State versus village-level governance, and decision-making in mixed arenas. Membership is open to all citizens (reg.133) and these inclusive Beach Management Units have authority, *inter alia*, to develop fisheries and fish landing station plans, to participate in selection processes for licenses in the relevant area, to collaborate with village councils on developing by-laws, and to settle fisheries disputes (reg.134).

Under the main Act, the declaration of protected areas and species also requires consultation with 'competent persons within the public and private sectors knowledgeable on environment issues' (s.23). The *Fisheries Regulations 2009* expand on these provisions, requiring the Director to consult with stakeholders in declaring critical habitats, and exempting hand line and non-vessel fishing from the Regulations (reg.54).[3] The Regulations also incorporate further opportunities for public participation. For example, the Director sets the duration of the prawn fishing season (reg.20), and prepares Sector Investment Guidelines (reg.26), in consultation with stakeholders. The Act makes other provisions for local level involvement in governance. Section 5(9), for example, notes that the Act does not prevent any local authority appointing relevantly qualified personnel to discharge its functions; and the Director must endeavor to ensure that 'local government authorities and associations of local authorities and other fisheries management authorities are consulted and kept informed' about fisheries laws and management plans (s.8(1)).

The Act establishes a Fisheries Development Fund whose purposes include promoting public awareness, through education and training, of the importance of sustainable development of fisheries resources, promoting and assisting 'the development of community management units by giving grants and assisting groups of persons wishing to form fisheries conservation and protection groups,' and assisting groups and individuals to participate in 'public debates and discussions on fisheries' and 'to ensure compliance with the Act' (ss.29–30).

In terms of access to justice, the *Fisheries Regulations 2009* states that any person aggrieved by a decision to revoke or suspend a license, or restrict it through conditions, can appeal to the Minister (reg.23). The Minister is required to respond to the appeal, in writing. Similarly, anyone aggrieved by a decision under the Regulations, can appeal to the Minister (reg.131).

### *Mauritius*

The relevant Mauritian statute is the *Fisheries and Marine Resources Act 2007*, which although relatively recently adopted, includes no specific objects or principles. The Act makes no reference to traditional or customary fishing rights, which is not unexpected given that the islands were uninhabited prior to European exploration and there are no Indigenous peoples' practices to safeguard. Yet whilst there are local communities who have fished for subsistence and livelihoods for a long time. As will be seen below, no special arrangements are made for these groups and only limited reference is made to artisanal fishers.

The Act does not establish any fisheries and marine resource institutions, but rather empowers the Minister to set up consultative committees which may be comprised of the Minister and 'other persons as the Minister may appoint' (s.3). This does not suggest that local communities or representatives from the small scale fishing sector would be involved. Fishery plans are not mandated as a management measure under the legislation, however vessels and fishermen must be registered, and areas and species may be protected. The Act appears to require registration of all fishing vessels, including those less than 12 m in length, with no exemption for subsistence fishers (ss.6 and 36), and all 'fishermen' including artisanal fishers (s.11). However, Sections 28 and 29 require gear licenses for specific net and gear types including gill, canard, large, pocket and shrimp nets, basket traps and bait gear. The combined effect of these sections appears to be that a small net, rod, and line, or hand lines may be used from a beach by those that do not consider themselves 'fishermen.' Aquaculture is also covered in the Act, and although explicit reference is not made to small scale endeavors, the Act does refer to *barachois*, which are traditional fish farms (s.8). Significantly, the Act requires authorization to establish a fish farming initiative, including a social impact assessment to determine the impact and any benefits accruing to local communities (s.8B (2)(b)(ii)).

The Act empowers the Minister to declare a protected area, by regulation, but the procedure and consultation requirements are not set out (s.4). The Act also seeks to protect fish and marine living resources at risk (s.16). Relevantly, the government can authorize the catching of a protected fish and/or collection of turtle eggs for purposes 'beneficial to the community' (s.16(2)(a)). This provision allows the government to have regard to cultural values, but does not empower communities to be involved in decision-making. Overall the Mauritian Act makes no provision for public participation in decision-making and fisheries management, but does allow the government to take supportive actions for local communities in limited circumstances.

The Act includes no broad right of standing to sue, nor any appeal provisions for a person aggrieved by a decision by the Minister.

### Seychelles

The Seychellois *Fisheries Act 2014* sets out its objects in Section 4, including 'effective management and sustainable development of fisheries,' through an ecosystem approach that does not jeopardize options for future generations 'to benefit from the full range of goods and services provided by marine ecosystems.' The statute is therefore underpinned by contemporary scientific approaches and environmental principles, though it does not reference inclusivity.

The Act vests in the Seychelles Fishing Authority the power to prepare and review management plans (s.5). The Authority itself is established under the *Seychelles Fishing Authority (Establishment) Act 1984*, which simply states that the President shall appoint five members of the Board, empowers the Authority to promote, organize and develop fishing and fishery resources, assist in policy formulations and, relevantly, conduct negotiations with regard to fishing or fisheries. There are no specific provisions that ensure public representation in the Authority or its functions.

In terms of fishing and/or access rights, the Act makes no specific provisions. As with Mauritius, this is not entirely unexpected given that the islands were originally uninhabited with no Indigenous peoples. Nonetheless, the Act does recognize local fishers in more explicit ways than neighboring Mauritius. Nevertheless, no power is expressly given to make regulations for small-scale or local fishers. There are no exemptions from vessel registration requirements (s.8), except those used for recreational fishing (s.21), and indeed the Act makes specific reference to local fishing vessels and Seychelles fishing vessels requiring licenses (s.16). The Minister is also empowered to make regulations for any type of fishing, with or without the use of a vessel which would also cover shore-based subsistence fishing (s.20).

The Act sets out the elements that may be included in a fishery management plan, including *inter alia* the 'role of stakeholders in decision making relating to the management plan' (s.5(f)). Furthermore, the Authority is required to 'consult the fisheries industry, local fishermen and such other persons engaged in fishing and fishing related activities as appear to the Authority to be appropriate' (s.5(3)), and if it considers necessary it 'may enter into a co-management arrangement with any person' (s.5(4)). Co-management is defined as including partnership arrangements with users, NGOs and other stakeholders (s.3). The plan itself, or the review of the plan, must be gazetted providing opportunities for the public to have access to its contents (s.5(7)). This is matched by requirements for fishers to provide fishing related information to the Authority for analysis (s.7). These provisions are inclusive and encourage stakeholder involvement in management.

In terms of access to justice, the Seychellois legislation establishes an Appeals Board whose membership is to include at least two members with 'knowledge and experience in [the] fishing industry' (s.38). Any person whose 'interests are adversely affected by an order, direction or other decision of the Authority, and who is dissatisfied with the decision, may appeal against the decision to the Appeals Board' on any matter which is 'manifestly unfair' or contrary to the Act (s.43). This appears to be a broad standing provision providing access to justice to all members of the affected public. Nevertheless, no appeal lies against 'a decision of the Minister about making a management plan or regulations for measures or plans for the management of fisheries,' thus weakening the ability of the public to appeal against the contents of a management plan that may affect their fishing rights or livelihoods (s.43(2)(c)).

## Comparative analysis

The above sections demonstrate that none of the jurisdictions have all the ideal elements of a strongly inclusive regulatory framework: inclusive institutions, broadly participative processes, and multiple rules to facilitate fishers' involvement in the management of fishery resources. By comparing instruments, though, best practice provisions can be drawn out. The most inclusive provisions of each of the different statutes together, can provide a foundation for the development of a 'model' fisheries instrument. This model could have, for example (1) objects, principles, and preservation of cultural practices and traditional fishing rights drawn from Kenya; (2) institutional membership along the lines provided for in Sri Lanka; (3) differentiation of scales and types of fishing operations expanding upon the approach taken in Bangladesh; (4) co-management opportunities developed from the Kenyan and Tanzanian examples and (5) access to justice safeguards similar to those in Seychelles.

Examining these areas in more detail, it can be seen that the Kenyan Act has clear and concise objects and underlying principles which provide a strong foundation for inclusive governance. The Kenyan legislation is also one of the few instruments that refer to youth and women, as well as alternative livelihoods. The Tanzanian statute specifically seeks to facilitate artisanal fishing and other community-scale marine resource livelihoods. Institutions created under Sri Lankan law have power to undertake social welfare and infrastructure, and activities of benefit to fishing communities. Customary fishing rights are also safeguarded under Kenyan legislation, and access to community fishing grounds under the Tanzanian law.

The Sri Lankan legislation is marked by its inclusive fisheries council. Two notable features are the number and naming of specific representative groups types. Only Kenya and Sri Lanka make specific reference to women and youth, even though these groups are known to dominate in specific small-scale fisheries in other countries: for example, sea cucumber fishery in Tanzania (Jiddawi & Öhman, 2002). Kenya has strong provisions to preserve and support livelihoods. Social impact assessments are also a feature of the Kenyan legislation, as well as the law in Mauritius.

Kenya, Tanzania, Seychelles, and Sri Lanka specifically provide for co-management; increasing opportunities for broad participation in local governance. Absent from the jurisdictions analyzed here is any example of complete devolution of authority through, for example, community-based management. Kenya comes closest to this with detailed regulations which allow for beach management unit to pass by-laws, albeit that they require central government approval. Yet local government, and the national fisheries Director, are still involved in management planning.

Relatively few of the instruments require broad public consultation on management plans, protected area designations, and species listing. Ideally, the broader public, beyond fishers and direct stakeholders, should have an opportunity to make submissions, and for them to be taken into account by decision-makers. Yet none of the statutes examined here automatically provide broad public rights. Kenya creates stakeholder consultation obligations and gives the decision-making body power to invite submissions. The Seychellois law includes consultation requirements in management planning. Instead, jurisdictions such as Bangladesh provide for the declaration of zones where only local fishing is permitted and/or other fishing restricted, but the central government makes those decisions. Seychelles also has useful provisions on marine protected areas.

Distinguishing between different types and scales of fishing operations is important, because licensing and approval obligations can provide a major regulatory hurdle for subsistence fishers and very small-scale operations. Bangladesh has sought to overcome this by exempting non-mechanized and local fishing vessels from licensing obligations, and also permitting areas to be declared where special rules can apply to such vessels. Tanzania has provisions which exempt traditional fishing methods from licensing requirements. Mauritian law permits the government to authorize the catching of protected marine animals for community benefit.

Access to information is a critical component, and necessary precursor to genuine involvement in decision-making. Again, the Kenyan legislation includes the most relevant provisions in this regard. Finally, access to justice ensures accountability and provides rights of redress. Seychelles provides the broadest rights of standing, with Kenya and Tanzania including more restricted access to justice provisions.

Largely missing from the legislation analyzed here is assistance to participate; only Tanzania makes specific reference to this. It has been recognized that assistance to participate is needed, including compensation for time lost on other tasks (Sobo, 2012). This is critical if widespread environmental stewardship is to be encouraged, and the necessary cultural shift achieved to secure sustainable fishery management.

## Conclusion

This article has demonstrated the importance of public participation in good fisheries governance, and explored the ways in which this can be achieved through an examination of fisheries regulations, institutions, and processes in six Indian Ocean States. There is little doubt that all jurisdictions embed some participatory approaches, yet none have strong provisions in all relevant areas, leaving room for improvement through law reform.

It is clear that further research is needed to identify a full suite of regulatory options which law-makers can use in improving fisheries governance in discreet jurisdictions. Furthermore, and equally critical, is the need for analysis of the law in action. How laws operate in reality is often very different from how they appear on paper. Therefore, empirical analysis is critical before law reform options can be identified. In the meantime, this article has highlighted the ways in which the public can participate, and stakeholders can be involved, in fisheries governance under existing Indian Ocean legal instruments. This research provides useful insights into opportunities for public participation in fisheries law in the region, and how the small-scale fishing sector is supported, as a basis for advancing best practice fisheries governance.

## Notes

1. In addition, there are several Indian Ocean dependent territories of the United Kingdom and France.
2. The Rio Declaration also emphasized the vital role women play in environmental management and encouraged their full participation in achieving sustainable development: Principle 20.
3. Marine protected areas are predominantly governed by the *Marine Parks and Reserves Act 1994*.

## Acknowledgements

The author would like to thank the peer reviewers and editor for comments and feedback on this paper, which have greatly improved it.

## Disclosure statement

No potential conflict of interest was reported by the author.

## ORCID

*Erika Techera* http://orcid.org/0000-0001-8283-2967

## References

Armitage, D. R., Plummer, R., Berkes, F., Arthur, R. I., Charles, A. T., Davidson-Hunt, I. J., ... Wollenberg, E. K. (2009). Adaptive co-management for social–ecological complexity. *Frontiers in Ecology & the Environment*, *7*(2), 95–102.

Arnstein, S. R. (1969). A Ladder of Citizen participation. *Journal of the American Planning Association*, *35*(4), 216–224.

Bavinck, M., Jentoft, S., Pascual-Fernández, J. J., & Marciniak, B. (2015). Interactive coastal governance: The role of pre-modern fisher organizations in improving governability. *Ocean & Coastal Management, 117*, 52–60.

Berkes, F. (2008). Evolution of co-management: Role of knowledge generation, bridging organizations and social learning. *Journal of Environmental Management, 90*, 1692–1702.

Bijma, J., Pörtner, H.-O., Yesson, C., & Rogers, A. D. (2013). Climate change and the oceans – what does the future hold? *Marine Pollution Bulletin, 74*, 495–505.

Blasiak, R., Spijkers, J., Tokunaga, K., Pittman, J., Yagi, N., & Österblom, H. (2017). Climate change and marine fisheries: Least developed countries top global index of vulnerability. *Plos One, 12(6)*, e0179632. doi:10.1371/journal.pone.0179632

Clifton, J., Etienne, M., Barnes, D. K. A., Barnes, R. S. K., Suggett, D. J., & Smith, D. J. (2012). Marine conservation policy in Seychelles: Current constraints and prospects for improvement. *Marine Policy, 36*, 823–831.

Coffey, C. (2005). What role for public participation in fisheries governance? In T. S. Gray (Ed.), *Participation in Fisheries Governance* (pp. 27–44). Copenhagen: Springer.

Cohen, P., et al. (2019). Securing a just space for small scale fisheries in the blue economy. *Frontiers in Marine Science*, doi:10.3389/fmars.2019.00171

Delgado-Serrano, M., Mistry, J., Matzdorf, B., & Leclerc, G. (2017). Community-based management of environmental challenges in Latin America and the Caribbean. *Ecology and Society, 22(1)*, 4–13.

Doyle, T. (2018). Blue Economy and the Indian Ocean Rim. *Journal of the Indian Ocean Region, 14*(*1*), 1–6.

FAO. (2015). *Voluntary guidelines for securing sustainable small-scale fisheries in the context of food security and poverty eradication*. Rome: FAO.

FAO. (2018a). *The State of World Fisheries and Aquaculture 2018 – meeting the sustainable development goals*. Rome: FAO.

FAO. (2018b). *Impacts of climate change on fisheries and aquaculture: Synthesis of current knowledge, adaptation and mitigation options*. Summary of the FAO Fisheries and Aquaculture Technical Paper 627. Rome: FAO.

Govt of Seychelles. (2017). *Seychelles' Blue Economy Strategic Policy Framework and Roadmap: Charting the Future (2018-2030)*. Retrieved from https://seymsp.com/wp-content/uploads/2018/05/CommonwealthSecretariat-12pp-RoadMap-Brochure.pdf

Haambiya, L., Kaunda, E., Likongwe, J., Kambewa, D., & Muyangali, K. (2015). Local-scale governance: A review of the Zambian approach to fisheries management. *Journal of Agricultural Science and Technology B, 5*(*2*), 81–92.

Hassan, S. M. D. (1999). Public participation in environmental law in Bangladesh. *Asia Pacific Journal of Environmental Law, 4*(*2*), 163–174.

IORA. (2017). *Blue Economy*. Retrieved from http://www.iora.int/en/priorities-focus-areas/blue-economy

IORA. (2019). *Member States*. Retrieved from https://www.iora.int/en/about/member-states

Jiddawi, N. S., & Öhman, M. C. (2002). Marine fisheries in Tanzania. *Ambio, 31(7/8)*, 518–527.

Kearney, J., Berkes, F., Charles, A., Pinkerton, E., & Wiber, M. (2007). The role of participatory governance and community-based management in Integrated Coastal and Ocean Management in Canada. *Coastal Management, 35(1)*, 79–104.

Lan, K.-W., Evans, K., & Lee, M.-A. (2013). Effects of climate variability on the distribution and fishing conditions of yellowfin tuna (*Thunnus albacares*) in the western Indian Ocean. *Climatic Change, 119*, 63–77.

Llewellyn, L.E., & English, S. (2016). A roadmap to a sustainable Indian Ocean blue economy. *Journal of the Indian Ocean Region, 12*(1), 52–66.

McGrath, C. (2010). *Does environmental law work? How to evaluate the effectiveness of an environmental legal system*. Saarbrücken: Lambert Academic Publishing.

Mohanty, S. K., Dash, P., Gupta, A., & Gaur, P. (2015). *Prospects of Blue Economy in the Indian Ocean*. RIS, India. Retrieved from http://www.ris.org.in/sites/default/files/Final_Blue_Economy_Report_2015-Website.pdf

Nairobi Statement of Intent on Advancing the Global Sustainable Blue Economy. (2018). Sustainable Blue Economy Conference 26-28 November 2018. Retrieved from http://www.blueeconomyconference.go.ke/wp-content/uploads/2018/11/Nairobi-Statement-of-Intent-Advancing-Global-Sustainable-Blue-Economy.pdf

Okello, N., Beevers, L., Douven, W., & Leentvaar, J. (2009). The doing and un-doing of public participation during environmental impact assessments in Kenya. *Impact Assessment and Project Appraisal, 27*(*3*), 217–226.

Pathmanandakumar, V. (2017). The Effectiveness of Co-management practices: The case of small-scale fisheries in Sri Lanka. *Journal of Aquatic Resource Development, 8*, 9. doi:10.4172/2155-9546.1000509

Singh, A. (2015). Climate change and maritime security in the Indian Ocean region. *Journal of Defence Studies, 9*(*1*), 63–82.

Snijman, P. (2011). Review of the legal frameworks in the ESA-IO region. SmartFish REPORT/RAPPORT: SF/2011/13. Retrieved from http://www.fao.org/3/a-az383e.pdf

Sobo, F. S. (2012). *Community Participation in Fisheries Management in Tanzania*. IIFET 2012 Tanzania Proceedings. Retrieved from https://ir.library.oregonstate.edu/concern/parent/1c18dg58c/conference_proceedings_or_journals/xd07gt68r

Stohr, C., Lundholm, C., Crona, B., & Chabay, I. (2014). Stakeholder participation and sustainable fisheries: An integrative framework for assessing adaptive comanagement processes. *Ecology and Society, 19*(*3*), 14–27.

Suárez de Vivero, J. L., Rodríguez Mateos, J. C., & Florido del Corral, D. (2008). The paradox of public participation in fisheries governance. The rising number of actors and the devolution process. *Marine Policy, 32*(*3*), 319–325.

Techera, E. (2018). Supporting blue economy agenda: Fisheries, food security and climate change in the Indian Ocean. *Journal of the Indian Ocean Region, 14*(*1*), 7–27.

Techera, E. (2019). Achieving Blue Economy goals: The need for improved legal frameworks across the Indian Ocean. *Seychelles Research Journal, 1*(*2*), 5–14.

UNCED. (1992). Agenda 21: Report of the UNCED, I, UN Doc/CONF.151/26/Rev.1. Retrieved from https://sustainabledevelopment.un.org/content/documents/Agenda21.pdf

UNDP. (2011). *Towards Human Resilience: Sustaining MDG Progress in an Age of economic Uncertainty.* Chapter 8 – Governance Principles, Institutional Capacity and Quality. New York: UNDP.

UNECOSOC. (2006). *Compendium of basic terminology in governance and public administration.* Committee of Experts on Public Administration. E/C.16/2006/4.

UNGA. (2015). *Transforming our world: the 2030 Agenda for Sustainable Development*. UNGA A/RES/70/1. Retrieved from http://www.un.org/ga/search/view_doc.asp?symbol=A/RES/70/1&Lang=E

van der Elst, R., Everett, B., Jiddawi, N., Mwatha, G., Afonso, P., & Boulle, D. (2005). Fish, fishers and fisheries of the Western Indian Ocean: Their diversity and status. A preliminary assessment. *Philosophical Transactions: Mathematical, Physical and Engineering Sciences, 363*(*1826*), 263–284.

Voyer, M., Schofield, C., Azmi, K., Warner, R., McIlgorm, A., & Quirk, G. (2018). Maritime security and the Blue Economy: Intersections and interdependencies in the Indian Ocean. *Journal of the Indian Ocean Region, 14*(*1*), 28–48.

Wickramasinghe, W. A. R., & Bavinck, M. (2015). Institutional landscapes affecting small-scale fishing in Southern Sri Lanka – legal pluralism and its socio-economic effects. *Maritime Studies, 14*, 18–36.

Wiser, G. M. (2000). *Transparency in 21st century fisheries management: Options for public participation to enhance conservation and management of international fish stocks.* New York: The World Wildlife Fund-US.

Yang, T., & Percival, R. V. (2009). The emergence of Global Environmental Law. *Ecology Law Quarterly, 36*(*3*), 615–664.

# Ecosocial work through corporate social responsibility: the case of company engagement with civil society in coastal communities in Mauritius

Brita Backlund Rambaree

**ABSTRACT**

This article presents a case study of corporate social responsibility (CSR) in coastal areas of Mauritius where company CSR work has brought new ways of engaging with communities. The study examines in what ways companies are involved in ecosocial work and what the challenges and possibilities are from an environmental justice perspective. Based on interviews with representatives of companies and other organizations involved with CSR the study shows that CSR involvement in ecosocial work may bring possibilities for advancing environmental justice for local communities as it can enhance resources and capacity building, as well as bring actors together and play a role in advocacy and empowerment work. Challenges include that companies may have difficulties in contributing to greater justice in terms of community involvement. Moreover, inequalities and substantial environmental justice issues, such as access to land and safe housing, are difficult to fully address through a CSR framework.

## Introduction

Corporate Social Responsibility (CSR) involves the merging of economic considerations with social and environmental aspects and it is concerned with how companies engage with those affected by their operations, such as local communities. As such the CSR work of companies bears relevance for local social and environmental development and how sustainability is achieved in local contexts. How specifically company CSR work relates to local communities and ecosocial work will be examined in this article based on a case study of CSR in the Mauritian tourism and international real estate sectors. Mauritius is a country where company CSR involvement has been made mandatory and where regulations direct company involvement in CSR towards socio-economic development and environmental issues. In this context much of the CSR involvement can be seen as activities in the field of social work since companies are involved in enhancing wellbeing and promoting social change and development through, for instance, setting up welfare services, organizing and providing resources for community work, and undertaking advocacy and political interventions. In part, and particularly in coastal areas of Mauritius, company CSR involvement is oriented towards addressing vulnerabilities that often concern the social as well as the natural environment. This entails companies both financing and being directly involved in interventions that could be labeled ecosocial work. The term ecosocial work refers to social work that combines social and ecological perspectives and takes a holistic approach to sustainable community development (Matthies, Närhi, & Ward, 2001; Närhi & Matthies, 2018). While this term is generally not used in connection with company CSR work, Mauritius provides an informative case on how CSR in fact may involve companies becoming key actors in local community development and ecosocial issues. However, an ecosocial work approach implies connecting vulnerability and marginalization to socio-economic structures, power imbalances in society and the upholding of injustices (Närhi & Matthies, 2018; Rambaree, Powers, & Smith, 2019). While such an approach may entail opposition to the activities of companies it can also be usefully applied to develop the way that companies interact with local communities on ecosocial issues. The values and perspectives attached to ecosocial work can then facilitate rethinking CSR engagement and taking CSR in the direction of greater positive contributions to the society.

As CSR globally has gained in importance as a framework for addressing social and environmental issues, yet faces criticism for lack of impact and significance (see for instance Banerjee, 2008; Blowfield & Murray, 2014; Fleming & Jones, 2013; Hanlon, 2008; Nyberg, Spicer, & Wright, 2013), it is important to consider how an ecosocial perspective can improve CSR involvement in local communities. This article therefore uses CSR involvement in coastal communities in Mauritius as an informative case of ecosocial work through CSR. It shows how a core value of ecosocial work, environmental justice, can be applied to assess the contributions of company community engagement. The article seeks to address two research questions: (a) In what ways are companies involved in ecosocial work in coastal communities through CSR?; (b) What are the challenges and possibilities of this involvement from an ecosocial work perspective? The paper presents the findings of interviews that were conducted with representatives of companies and other organizations involved with CSR and provides an analysis of CSR involvement in terms of environmental justice.

The article is structured as follows: First a background is provided of CSR in Mauritius. Thereafter the article outlines the theoretical framework of environmental justice. This is followed by a presentation of the methods and then the analysis. The paper ends with a discussion and conclusion

## Background: CSR in Mauritius

Mauritius is an island state in the Indian Ocean with a population of 1.3 million. The country gained independence from the United Kingdom in 1968 and has since experienced rapid economic growth. It has developed from a low-income economy dependent on the revenue from sugar cane to a diversified upper middle-income economy considered one of the strongest and most stable economies in Africa (Brooks et al., 2017; Svirydzenka & Petri, 2017). However, many social challenges remain to be addressed and segments of the population are experiencing increasing vulnerability and social exclusion (Tandrayen-Ragoobur & Kasseeah, 2018). Government efforts to address the social challenges have, since 2009, used the CSR framework to pool company resources into promoting socio-economic development. This has meant that CSR has become regulated and largely directed by the government. The regulations require all profitable companies to contribute two percent of the chargeable income as CSR contributions. From the onset the funds were directed at social problems and issues related to socioeconomic development with the initial regulations stating that companies should spend the funds 'on CSR activities approved by Government or to transfer these funds to Government to be used in the fight against poverty' (National Empowerment Fund, 2011). Many large companies chose to set up their own CSR foundations that initiated projects related to social and environmental issues, often in areas where the companies have operations. These foundations remain an important part of how companies work with CSR and they are active in community development and engage with civil society at local level. However, in 2017 a National CSR (NCSR) Foundation was set up to enhance coordination and transparency and to improve the outcomes for the beneficiaries (National CSR Foundation, 2017).[1] This meant that companies had less choice in how to allocate their two percent CSR contribution: At least 50 percent of each company's CSR contributions had to be channeled through the NCSR Foundation. As from 2019, this has been increased to at least 75 percent of the CSR contribution (Income Tax Act 1995, Consolidated Version). The NCSR Foundation is governed by a council where the government as well as the private sector and NGOs are represented. The foundation uses the company contributions to fund projects by NGOs for the benefit of vulnerable groups. These projects should fall within one or more of ten priority areas: socioeconomic development as a means for poverty alleviation; educational support and training; social housing; supporting people with disabilities; dealing with health problems; family protection including gender-based violence; leisure and sports; environment and sustainable development; peace and nation-building; and road safety and security (National CSR Foundation, 2017). The part of a company's CSR contribution that is not channeled through the NCSR Foundation should be used for implementing a CSR program or funding a CSR program provided by an NGO. Also here the funding should go to the priority areas of intervention. Thus, the CSR framework in Mauritius has become to a large extent state-directed but entails the private sector being involved through funding, steering and implementation of projects and programs for the benefit of local communities.

In addition, the private sector in Mauritius is also involved in local level sustainability and socio-economic development through requirements that have been placed on the real estate sector. The real estate sector has experienced rapid growth in the last few years, facilitated by government efforts to attract foreign direct investment by enabling non-citizens to purchase residential property in coastal areas. Initially, the inflow of foreign direct investment through the real estate sector was promoted through the Integrated Resort Scheme (IRS), which facilitated conversion of large areas of coastal land into property projects consisting of upmarket villas and apartments. This was later followed by the Real Estate Scheme (RES) that enabled the conversion of smaller portions of land into property projects. Both schemes have, as from 2017, been merged into the Property Development Scheme (PDS). With the PDS the government introduced regulations intended to spread the benefits of the property projects also to local communities. The intention has been for the PDS 'to ensure that the benefits of the development of the Scheme accrue to neighboring communities and to small entrepreneurs generally' (Investment Promotion (Property Development Scheme) Regulations 2015). Thus, each PDS project has to be preceded by a social impact assessment and a social needs assessment focusing on the local community. In addition each PDS project has to have a social fund for implementing 'a social programme identified in the social needs assessment and provide for social amenities, community development and other facilities for the benefit of the neighbouring community where the PDS Project is implemented' (Investment Promotion (Property Development Scheme) Regulations 2015). The social fund is constituted from a levy of approximately EUR 5000 per each residential unit that the PDS company has sold. Given the social element of the PDS it can be seen as a form of socially responsible enterprising or in other words a form of CSR. Consequently, in this article both the CSR scheme and the PDS scheme will be considered in terms of company CSR engagement with local communities.

In coastal areas both CSR and the PDS scheme entail companies being important actors engaging with the civil society and being involved in promoting socio-economic development and addressing social and environmental issues. Coastal areas in Mauritius are faced with numerous challenges including large inequalities, unemployment, low education levels, shortage of land, squatter dwelling, beach erosion, coral bleaching and pollution. Hotel companies and PDS companies have come to play an important role in addressing these challenges and often these companies are involved, through financing and direct interventions, in ecosocial work in the coastal areas where they operate. Their work bears relevance for local social and environmental development and it is therefore important to examine the challenges and possibilities of this new form of community engagement. Can community engagement as it is promoted by these companies advance the transition to a more sustainable society?

## Ecosocial work and environmental justice

Ecosocial work can be seen as social work that combines social and ecological aspects and takes the two as interlinked (Närhi & Matthies, 2018). While different perspectives exist on ecosocial work, a key element in ecosocial work is how ecosocial issues connect to vulnerability, inequalities, economic structures and corporate decision-making (Närhi & Matthies, 2018; Rambaree et al., 2019). To approach such structural aspects of ecosocial work it is useful to draw on the discourse of environmental justice. The discourse of environmental justice was initially used in discussions concerning inequities in exposure to environmental risks but has expanded to cover many more issues and new theoretical conceptualizations of the term (Schlosberg, 2013). As such, studies using an environmental justice perspective focus on, for instance, how socially disadvantaged communities are more exposed to environmental risks (Campbell, Greenberg, Mankikar, & Ross, 2016), how communities are vulnerable to the actions of companies (Ross, 2013), how environmental degradation affects particularly poor people (Dominelli, 2012), how marginalized communities are affected by climate change (Hetherington & Boddy, 2013), and how damage to the ecosystem has led to increased vulnerabilities (Ross & Zepeda, 2011).

The research presented in this study draws particularly on Bell's (2014) work on environmental justice. Bell suggests that environmental justice can be seen to incorporate, *distributive, substantive* and *procedural justice*. This framework is here applied to the case of company involvement in coastal communities in Mauritius.

*Distributive environmental justice* is defined by Bell (2014, p. 22) as 'an equitable distribution of environmental "goods" and protection from environmental harms for all socioeconomic groups'. This aspect of environmental justice covers inequities in relation to different social categories such as income, class or gender and it also covers equity in relation to different outcomes, whether in terms of resources, primary goods, capabilities, welfare outcomes or environmental risks (Bell, 2014). Further, distributive environmental justice encompasses equity in the distribution of both environmental harms and environmental 'goods' such as water, clean air and energy (Bell, 2014) Distributive justice aspects are often what has been the focus of environmental justice research (Bell, 2014; see also Schlosberg, 2013). However, Bell (2014) argues that it is not sufficient to only consider these aspects and that the environmental justice discourse should also embrace substantive and procedural justice as essential components.

*Substantive environmental justice* is defined by Bell (2014, p. 22) as 'a healthy environment'. While 'a healthy environment' can be defined in many ways, and even regulations vary on the environmental 'goods' and the hazards that affect individual and population health, it is an important aspect of environmental justice since it is the basis of human survival (Bell, 2014). In addition, procedural environmental justice is also required, according to Bell's conceptualization. *Procedural environmental justice* is concerned with 'fair, participatory and inclusive structures and processes of environmental decision making' (Bell, 2014, p. 22). As Bell (2014) outlines, there are different views on the possibility of achieving such structures and processes, but ensuring that they are in place is nonetheless necessary for environmental justice.

Drawing on Bell's (2014) working definition of environmental justice it is possible to specify what environmental justice would imply in the context of company-funded ecosocial work with local communities. Bell (2014) suggests a set of indicators that can be used for assessing the environmental justice situation in terms of distributive, substantive and procedural justice aspects. As such, indicators of *distributive environmental justice* include for instance equal access to 'goods' such as safe drinking water, sanitation, adequate and safe transport, and green space for recreation and leisure. They also include equal access to adequate housing and safe working and living environments, as well as equal protection from environmental disruptions and hazardous substances. As indicators of *substantive environmental justice* Bell (2014) includes for instance universal access to environmental 'goods' such as safe drinking water, sanitation, adequate and safe transport and green spaces, but also that everyone has adequate housing and are protected from hazardous substances. *Procedural environmental justice* is taken to include, for instance, that all parties affected by decisions are included in decision-making, received accurate and accessible information, and are able to draw on sufficient skills and personal resources to participate on an equal basis (Bell, 2014).

## Methods

The study was undertaken using a qualitative deductive approach where the theoretical perspective of environmental justice was applied to interviews on the practice of ecosocial work in coastal communities. A total of six interviews provided the main source of data. In addition, reports and other published materials on CSR in coastal communities were used in the process of preparing for the interviews. The interviewees were selected through purposive sampling to represent different perspectives on CSR issues in Mauritius. The interviewees were informed about the purpose of the study and their approval was sought before recording the interviews.

**Table 1.** Sample overview.

| No. | Interviewee position | Mode of conducting interview |
|---|---|---|
| IP 1 | CSR Manager | Via Skype |
| IP 2 | CSR Manager | By phone |
| IP 3 | Leader of a social movement | By phone |
| IP 4 | CSR Manager | Via Skype |
| IP 5 | Leader of an NGO | Face-to-face at the office of the interviewee |
| IP 6 | State-employed social worker working with CSR | By phone |

Table 1 gives an overview of the sample. The CSR managers represent three different large companies all of which operate hotels and/or PDS projects in several locations along the coasts of the main island of Mauritius. The social movement is involved in eco-social issues in coastal areas across Mauritius and the NGO has activities with a nationwide coverage. In total the interviewees included three women and three men. All interviews were carried out during 2018 and most were conducted by phone and Skype. This approach was chosen to fit the interviews around the busy schedule of the interviewees. The interviews lasted between 30 and 90 min and were recorded.

The recorded interviews were analyzed with the help of Atlas-ti software, version 7.5. Theoretical thematic analysis was used as the method for analyzing the data and this involved identifying themes and patterns in relation to the research questions and the theoretical framework (Braun & Clarke, 2006, 2013). Initially the entire dataset was coded by the author. The coding focused on searching for, categorizing and memoing types of interventions, meanings given to these as well as challenges and possibilities, seen from an ecosocial work perspective and assessed in terms of environmental justice. The codes were then combined into themes and the themes were checked for consistency against the coded extracts and the entire data set (Braun & Clarke, 2006). In the presentation of the findings, the names of persons and organizations were removed so that the interviewees will not be easily identified.

Given the small sample size the findings may only give an indication of company involvement in ecosocial work in Mauritius and do not necessarily reflect how other companies and organizations work with CSR and ecosocial work. Another limitation of the study concerns gender in relation to environmental justice. How environmental justice is experienced may differ greatly between men and women in a community (see for instance Stein, 2004) but the small sample size did not allow any conclusions about this.

The following section presents the analysis of the interviews. The analysis applies an environmental justice perspective – a core value of ecosocial work - in analyzing the contributions of company community engagement. The analysis is divided into two parts that respond to the research questions: (a) In what ways are companies involved in ecosocial work in coastal communities through CSR?; (b) What are the challenges and possibilities of this involvement from an ecosocial work perspective?

## Findings and analysis

### *Company CSR involvement in coastal communities*

CSR in the coastal communities is oriented towards addressing several pertinent social and environmental issues. Much of the involvement consists of the companies funding programs and projects, but the companies are also more directly involved in provision of services and in interacting with the local communities and NGOs, as well as promoting forums for such interaction.

### *Addressing unemployment and low levels of education*

CSR in the coastal communities is partly concerned with training and capacity building related to employment, for instance employability training and training focused on learning an occupation. Other central aspects of the CSR-work focused on promoting entrepreneurship and small businesses. This orientation of the CSR involvement has to be understood in the context of the high levels of unemployment and low levels of education in coastal areas of Mauritius. The three CSR managers (IP 1, IP2 and IP4) explained the rationale for this form of involvement. As IP 1 explained: 'Work was the issue (...) and the education level was and still is very low. It was always like this far out island with only one road and I must say that the authorities didn't pay much attention.' IP 4 also felt that unemployment is a main issue in the local area:

> Since they didn't have a good job or enough revenue when they had children they also they couldn't educate their children and it's a vicious cycle. The children also were not educated so they have these little jobs with little revenue. It's like continuous, generation after generation they remain poor. (IP4)

The training programs that are offered focus on employment in the hospitality industry and thereby provide opportunities for employment in the hotels and villas in the locality. As IP 2 pointed out 'In this region it was clear that one of the main possibilities for employment was hospitality'.

For the companies this type of CSR involvement is also serving the purpose of supplying suitable candidates for employment: 'It is also in their interest to get workers from next door; less transport, less cost [...] and in the contact tourists ask a lot about the locals so the more you have locals the better it is' (IP 1). The focus on employment can also help the companies to benefit from better trained and qualified workers while assisting people in the local community. IP 2 explained that apart from providing training for employment in the hospitality sector, the company also started a program for training and licensing of persons working as skippers on boats that operate in the tourism sector: 'They were skippers but they did not have any license to operate. [...] This has improved their living conditions because as they were illegal skippers they were not well paid.'

Also the promotion of entrepreneurship and small-businesses are forms of CSR involvement that respond to the needs of the hospitality industry while assisting the local community:

> There is no market place, there are four villages but there is not even one market place so I'm going to build a market place to try and stimulate exchange. Because villa owners are willing to buy locally [...] but there is no place for them to come to. And some of them [members of the local community] could sell those products that they are producing. We just need to give them the facility for it. (IP 1)

Undertaking CSR activities focusing on employment and entrepreneurship connected to the hospitality industry also means that the companies can draw on the expertise that they have among their employees: 'There are different skills that we already have in the hospitality sector, and experts so they provide the training to these people. [...] What we try to do is to make that person become employable ... ' (IP 4)

Basic skills training is also a part of the CSR activities and an activity that companies engage in to promote employment in areas with low education levels:

> We had to also start with life skills management and literacy and numeracy for adults. [...] One of the success stories, I can say, was a lady that has been able to set up a micro-enterprise because now she is able to use a calculator, she can read and communicate and interact with clients. (IP 2)

At the level of individuals in the local community, the involvement of companies in training and capacity building for employment and entrepreneurship translates mainly into contributions to social enhancement; but their involvement can also be seen as actions to promote sustainability in the local communities where the companies operate. To some extent, promoting sustainability is also focused on internal CSR aspects as a way of impacting on local communities: 'buy local, support local farmers, [...] fair trade, training of our own staff on how they need to manage their own resource consumption and in turn they take it to their home.' (IP 4)

### *Addressing issues concerning natural and physical environment*

The social issues experienced in the community are connected to the locality and any effort to promote community development would need to consider social issues as well as environmental issues. Some of the CSR involvement was initiated after social needs assessments that preceded property projects and, as such, it aims to broaden the benefits of property projects to the civil society. Involvement more directly oriented towards environmental issues was also part of CSR activities in the coastal communities, at times connected to training for employment: 'we also include in the training programme [on basic skills] a module on marine protection' (IP 2). Environmental education forms part of CSR, and other activities include raising awareness on protection of the lagoons, both among local residents and tourists, and organizing beach cleaning.

One of the interviewees (IP 2) also explained that the company is involved in improving sanitary facilities in the local community, in addressing the problem of asbestos in local houses and in addressing the need for social housing.

> A main project that we achieved [as member and financier of a collaborative community association] was to build sanitary facilities for the people [in the locality]. For years these people they did not have any sanitary facilities, I mean toilet and shower [...] We can have five families in a very small plot of land so there is no space [...] what we did is to have in one plot of land at least one shower [...] It was at least one thing that we could do.

### *Advocating for change*

In the same community another problem is asbestos, which exists in many of the houses in the area that were built by government after a cyclone in the 1960s. 'Unfortunately up to now there is no attempt to address it [...] there has been no single asbestos dismantling programme' (IP 2). The CSR manager felt that the government has to develop a dismantling plan and the company is therefore involved in advocacy on the issue: 'We are still advocating, we are still having discussions with the government' (IP 2). Dismantling the houses would require access to other housing for the residents and this is currently not available as there is a shortage of low-cost and social housing in the area, as well as a shortage of land: 'Social housing is another very big issue in this region. For more than 50 years there has been no social housing project in the region' (IP 2). Through their involvement in a collaborative community association the company is advocating for the provision of social housing, creating awareness on the issue and mobilizing the community in demanding for change. The company and the community association are putting forward the argument that the social fund provisions of new PDS should be primarily allocated to buying land from private land owners in the area so that the government can construct new social housing.

Advocacy and social mobilization were also part of company CSR activities in relation to other issues. IP 1 explained that their current CSR involvement involves focus on advocacy work.

> There are a few associations now, there wasn't then, like 15, 20 years ago [...] but they don't really know how to advocate, obtaining better infrastructure from the authorities. Something simple: transport is terrible [...] but they don't know how to get that done.

The same interviewee (IP 1) explained that their CSR efforts have involved helping the local community to advocate for change:

> We've stimulated a lot the CSOs, community service organisations, who are now advocating, I literally train them in advocacy, I gave them teaching in advocacy so they could know how to gather information, how to prepare a strategy to get things for people on the field.

### *Addressing gaps in service provision*

In addition to such interventions, company involvement is also concerned with direct provision of assistance and services to individuals in the local community. IP 1 was in the process of setting up a center with two social workers in one of the coastal villages. The center is intended to connect local residents to services and resources and provide training for employment. IP 2 has also, through the company's involvement in a collaborative community association, been involved in the setting up of a social center where counseling is offered, as well as training, awareness campaigns and prevention programs on alcohol and substance abuse, and music sessions.

## *The challenges of promoting ecosocial work in coastal communities through CSR*

### *Substantive environmental justice*

Substantive environmental justice entails a healthy environment for all with universal access to environmental 'goods', as well as adequate housing for all and protection from hazardous substances (Bell, 2014). While company CSR involvement in coastal communities in Mauritius involves a wide range of efforts to address social and environmental issues, it is apparent that the CSR framework does not fully address substantial issues such as shortage of land, crowded housing and the dismantling of houses with asbestos. IP 2 pointed out that the private sector can be part of addressing such issues but felt that the government has to get more involved: 'Social housing is not the responsibility of the private sector. Private sector of course can help [...] It is not because we talk about

CSR that we have to substitute the responsibility of the government'. On the issue of asbestos IP 2 also called for more government involvement: 'I'm not saying the private sector should not be involved, but the government should drive it'. Companies, such as the one that IP 2 represents, may have a role in drawing attention to these problems and in this way bring the local community closer to achieving substantial environmental justice. The same may be the case with other issues, such as public transport, where companies can assist communities in demanding improvement. However, CSR on its own does not solve these issues.

### *Distributive environmental justice*

Distributive environmental justice concerns equal access to environmental 'goods' and equal protection from environmental hazards. Companies in coastal communities of Mauritius contribute in important ways to improve the situation for the villagers but the inequalities that exist in these communities are often difficult to address through CSR. An important part of CSR in these communities is for instance company involvement in training and capacity building related to employment in the hospitality industry. This entails training for relatively low-paid jobs and may do little to change the unjust distribution in coastal communities and the inequality in access to land and housing. As waiters, gardeners and housekeepers many villagers obtain employment and an income but often would not be able to afford better housing and remain marginalized in communities that have increasingly become segregated with the upsurge of gated communities for the well-off. IP 3, representing a social movement working on issues related to coastal communities, was critical of the direction of development in coastal areas:

> ... you've got lots of people that are descendants of slaves and these guys are more and more feeling completely forgotten. You've got these huge millionaire villas popping up everywhere and they are being trapped in poverty. [...] You have clearly, a society going increasingly with de facto segregation and exclusion [...] The supply of social housing is woeful. The priority is on millionaire, expat villas ... We are going straight towards a social explosion on the coast. (IP 3)

The risk is also that some companies can exacerbate problems in coastal communities by purchasing or leasing land for property schemes or hotels. Large corporate developments may create jobs and revenue opportunities but at the same time they restrict how environmental 'goods' are shared and how opportunities to gain an income from tourism are distributed:

> A beach for a poor coastal village is their ticket out of poverty, it's their asset. That's what enables people to set up their SMEs, their little business in restaurants, rental of accommodation, guest houses, rental vehicles, bring people to do tours ... That's how people can work their way out of poverty. (IP 3)

As IP 3 pointed out, large corporate developments on the beachfront, restrict opportunities for the local community.

From an ecosocial work perspective another challenge is that CSR initiatives and funding cannot comprehensively cover everyone in need.

> A company "would never ever have enough CSR funds to fund the charities to actually do the work that should be done nationwide. A company's main focus is to make profit. It's a business, so they are not really here to look after the community. What they can do

> is try to improve [...] It can't become as a main focus. I mean for community development and social development, there is something called the state, it's the state's duty ... " (IP 3)

### *Procedural environmental justice*

Looking at procedural environmental justice further points to the challenges of doing ecosocial work through CSR. Procedural environmental justice concerns 'the fairness and transparency of the processes by which decisions are made' (Bell, 2014, p. 19). When companies promote and undertake ecosocial work activities they might be seen by themselves and others as the providers. An ecosocial work approach would instead build in participatory approaches where power is shared. From this perspective it is clear that when the company is the provider the civil society are the beneficiaries and the two would not be working to address issues on an equal basis. This can mean that companies plan and undertake activities to which they invite civil society organizations and members. The company would then set the agenda and the civil society has restricted involvement in the decision process. From the perspective of the 'provider' it can then be challenging to get the community to take the opportunities provided and to get appreciation for the work that is put in.

> ... it needs to be a two-way traffic we cannot force people, you know people need to help themselves. We cannot force. We can propose solutions but in the end we should not be taken for granted. Nobody should be taken for granted. There needs to be gratitude ... (IP 4)

When the company is seen as the provider the local community is disempowered:

> it's not good to do it in their place because they will think private sector can do it all. You have to understand the sugar industry world. The big boss was the boss, he was the king of the area, and they still ... they are disempowered. [...] When I go there they ask me things, not understanding that I can't obtain that for them. I have a role, no, for them I should be the one giving the permits, giving this, giving that. I'm the provider, so it's a process getting them to stop thinking that way and understanding they've got the power and it has to come from them. (IP 1)

Another aspect of procedural environmental justice relates to how company CSR funds are allocated. With the new CSR regulations 75 percent of company CSR funds have to be channeled through the NCSR foundation and the companies are then less free to allocate the funding by themselves. While the respondents expressed that this can enable a more systematic and transparent approach to allocating CSR funds, fears were also raised about the reduction in influence of NGOs and companies in the decision making. IP 5, leader of an NGO, saw benefits in having the NCSR Foundation: 'I think it is a good move to create the NCSR Foundation', but also explained that the more bureaucratic procedures have impacted on their ability to obtain funding:

> Since they were created they took two years to make one disbursement. So what do you do in the meantime? Now we have used the funding they gave us. What do you do? You sit and wait? Do I sack all these people? (IP 5)

Particularly at a transition phase, NGOs that have previously worked in close collaboration with companies may experience that they have less influence in the decision making. However, over a longer period of time, the new regulations can enable equal access to information and funding possibilities for all NGOs and hence greater procedural environmental justice. The centralization can provide better opportunities to include also smaller NGOs but, as IP 6 who works with CSR funding explained, this can be challenging:

> We have to train these people because we have very small NGOs who are coming now. [...]You know they don't have the know-how, they don't have the experience. [...] Before giving the money we have to train them. (IP 6)

### *The possibilities of promoting ecosocial work in coastal communities through CSR*

The interviewees all saw possibilities and benefits in using CSR to promote social development and address environmental issues in coastal communities. The possibilities and benefits are mainly in terms of developing *procedural environmental justice*. Through CSR it is, for instance, possible to improve resources and techniques among NGOs and hence enhance the ability of such organizations to take part in decision making and have the possibility to influence decisions concerning the local community. Through direct financial contributions from companies and through the funding channeled via the NCSR Foundation, the CSR framework benefits NGOs financially. Their contact with companies has also meant possibilities for capacity building:

> Since 2009 there's been a huge improvement in governance, and strategic thinking, you know, vision, mission, values, [...] We have not only given money, we have given a lot of resources in the way of people to the civil society. [...] When we were having training I would bring a few of the NGOs in the private sector training, so they could learn the techniques. [...] Now they can compete and ask for money internationally, for example EU. (IP 1)

CSR can also help local organizations to get the opportunity to voice their concerns, to do advocacy and to be empowered to work for change:

> Now that the people are under [name of collaborative community association] they are well aware, they are really conscious of the problem [...] Last week they had a press conference and they said that listen we are fed up with this problem [...] and we were not there, we left them to speak to the press because we want the community to come forward. (IP 2)

The CSR involvement of companies was also perceived to assist communities by bringing actors together to collaborate and work for change:

> We put in place a platform which regroups all the major stakeholders within the catchment area [...] All the major stakeholders were on the platform and we had communication, interactions, about the project, about the situation in the region. We were a driving force in this respect. (IP 2)

Indirectly, and over a longer period of time there is then also a possibility for CSR to contribute towards greater *distributive* and *substantial environmental justice*, since the CSR involvement assists communities to raise concerns and work collaboratively to address these concerns. As Bell (2014) points out, better environmental conditions depend on citizens being able to obtain information as well as organizing and stating their preferences.

Overall, much of the CSR involvement is long-term involvement and the company representatives saw their work as bringing lasting benefits to the community. For instance, IP 2 explained that the company, together with other corporate actors and a number of civil society organizations set up the collaborative community association as a continuation of what was initiated under a PDS social fund: 'Once we will sell all the properties there will be no more funds, so what we thought of was a strategy that will pursue what we started and in this context we launched the [collaborative community association]'. The association is today an NGO that has 'its own strategy and programme' (IP 2). It is funded through CSR contributions and private donations. IP 4 also shared the perspective that company involvement is a long-term undertaking:

> We are not just like [doing] philanthropy, going to give some money and then that's it, no. What we try to do is to make that person become employable and empower the person so that he can stand on his own feet [...] It's more long-term ...

The possibilities of relying on CSR for promoting ecosocial work are however dependent on how CSR is used in connection with government policies and interventions and the involvement of other actors in society. The private sector can provide important resources and capacity but as the interviewees pointed out it is important to further develop collaboration and partnerships. IP 3 argued for a modified framework that involves closer collaboration between the state, the private sector and the civil society:

> You can have a system whereby CSR can fund social innovation, can fund interesting programmes. A company can set up an excellent project, but the idea is that if you do something excellent then it should be replicated across the country. No company can have that kind of resources, it's up to the state so we should be having some kind of healthy partnership where the NGOs, charities come up with very innovative programmes, these are funded by the CSR. If it works great the state can take over and replicate it or fund a charity to do it nationwide. If it doesn't, well it's only CSR money by company X, Y, Z that has been lost. [...] Risk is taken by private sector CSR and state therefore invest taxpayers' money in successful programmes.

## Discussion and conclusions

This study has examined in what ways companies are involved in ecosocial work in coastal communities in Mauritius and what the challenges and possibilities are from an environmental justice perspective. Taken as a whole, the CSR involvement of companies in coastal regions of Mauritius encompasses both social and environmental issues and entails a broad approach to addressing issues and improving conditions for people in the local area. In this way the CSR involvement has some similarities with the approach to community work that was developed by Jane Addams in the early days of social work. Similarly to the community work undertaken by Addams, it entails efforts to aid by seeing the social environment as well as the physical and built environment (see Addams, 1910). An important difference however lies in the profit-interest, which by necessity would be reflected in CSR interventions. A company is a for-profit entity with responsibility also towards shareholders in terms of making a profit. A risk with CSR is then that it may aid the local community in some ways but at the same time do little to change the unjust distribution and prevailing inequalities in a society. As Ross (2013, p. 196) points out, 'From an anti-oppressive social work perspective, CSR reinforces "relations of domination and subservience" under the guise of "doing good"'.

However, to simply oppose company efforts on this basis would be to reject an opportunity to collaboratively work towards greater sustainability and enhanced well-being. McKinnon and Alston (2016) argue that a collaborative approach between social workers and other actors can make a difference in addressing issues of sustainability. As is evident from this study, companies can be important contributors, not least in terms of resources. It is then important to look critically at the possibilities and challenges of working with these issues through CSR. Applying an environmental justice perspective can be useful in this endeavor. As a core value of ecosocial work, environmental justice points to the importance of seeing the structural aspects of vulnerability, marginalization and ecosocial issues (Närhi & Matthies, 2018; Rambaree et al., 2019). These are aspects that have to be taken into consideration if we are to use CSR in a way that goes beyond the current limitations of how companies engage with communities.

The findings from this case study suggest that CSR involvement in ecosocial work brings possibilities particularly in terms of advancing procedural environmental justice for local communities as it brings resources, capacity building, helps to bring actors together and plays a role in advocacy and empowerment work. There are however also challenges in term of procedural environmental justice since companies may get the role as provider and thus face challenges in involving civil society. In terms of substantive and distributive environmental justice the challenges seem to currently outweigh the possibilities. Inequalities and substantial issues affecting the coastal communities are difficult to address through CSR, at least without the involvement of other non-corporate forms of intervention and active involvement of the civil society. To extend on the possibilities and benefits of company involvement in ecosocial work it is necessary to see how CSR can move in the direction of also promoting substantive and distributive environmental justice. That however requires more emphasis on a collaborative approach to CSR, as stated below:

> If we think that CSR is going to be solving all the social and environmental problems in the coastal community, well that's dishonest. It can only be part of a solution that has to be driven by the state in proper consultation with all the stakeholders. (IP 3)

## Note

1. After this study was carried out the NCSR Foundation has changed name to National Social Inclusion Foundation (NSIF).

## Acknowledgements

I am grateful to the interviewees for taking part in the study and to management at the NCSR Foundation (recently renamed NSIF) for providing useful background information. I would also like to thank the anonymous reviewers for constructive comments, as well as the guest editor of this special issue Benito Cao and the commissioning editor of the journal Adela Alfonsi for their quick and helpful feedback.

## Disclosure statement

No potential conflict of interest was reported by the author.

## ORCID

*Brita Backlund Rambaree* http://orcid.org/0000-0002-7094-7297

## References

Addams, J. (1910). *Twenty years at Hull House, with autobiographical notes*. New York: The MacMillan Company. 1912 (c.1910). Retrieved from https://digital.library.upenn.edu/women/addams/hullhouse/hullhouse.html

Banerjee, S. B. (2008). Corporate social responsibility: The good, the bad and the ugly. *Critical Sociology*, *34*(1), 51–79.

Bell, K. (2014). *Achieving environmental justice: A cross-national analysis*. Bristol: Policy Press.

Blowfield, M., & Murray, A. (2014). *Corporate responsibility* (3rd ed.). Oxford: Oxford University Press.

Braun, V., & Clarke, V. (2006). Using thematic analysis in psychology. *Qualitative Research in Psychology*, *3*(2), 77–101. doi:10.1191/1478088706qp063oa

Braun, V., & Clarke, V. (2013). *Successful qualitative research: A practical guide for beginners*. London: Sage.

Brooks, C., Lamport, M., Padachi, K., Sannassee, V., Seetah, K., & Seetanah, B. (2017). The impact of foreign real estate investment on land prices: Evidence from Mauritius. *Review of Development Economics*, *21*(4), e131–e146. doi:10.1111/rode.12316

Campbell, C., Greenberg, R., Mankikar, D., & Ross, R. D. (2016). A case study of environmental injustice: The failure in Flint. *International Journal of Environmental Research and Public Health*, *13*(10), 951–961. doi:10.3390/ijerph13100951

Dominelli, L. (2012). *Green social work: From environmental crises to environmental justice*. Malden, MA: Polity Press.

Fleming, P., & Jones, M. T. (2013). *The end of corporate social responsibility: Crisis and critique*. London: Sage.

Hanlon, G. (2008). Rethinking corporate social responsibility and the role of the firm: On the denial of politics. In A. Crane, A. McWilliams, D. Matten, J. Moon, & D. Siegel (Eds.), *The Oxford handbook of corporate social responsibility* (pp. 156–172). Oxford: Oxford University Press.

Hetherington, T., & Boddy, J. (2013). Ecosocial work with marginalized populations: Time for action on climate change. In M. Gray, J. Coates, & T. Hetherington (Eds.), *Environmental social work* (pp. 46–61). London: Routledge.

*Income Tax Act*. (1995). *Consolidated version*, Sub-Part AD. Retrieved from http://www.mra.mu/download/ITAConsolidated.pdf

*Investment Promotion (Property Development Scheme) Regulations*. 2015. Retrieved from http://registrar.mof.govmu.org/English/Documents/7.%20Investment%20Promotion%20Act%20Regulations(%20PDS).pdf

Matthies, A.-L., Närhi, K., & Ward, D. (Eds.). (2001). *The eco-social approach in social work*. Jyväskylä: Sophi Publishers.

McKinnon, J., & Alston, M. (2016). Introduction. In J. McKinnon & M. Alston (Eds.), *Ecological social work: Towards sustainability* (pp. 1–18). London: Palgrave Macmillan.

Närhi, K., & Matthies, A.-L. (2018). The ecosocial approach in social work as a framework for structural social work. *International Social Work*, *61*(4), 490–502.

National CSR Foundation. (2017). Strategic plan 2017–2019: Towards a new CSR framework for collective action on social inclusion and equity. Retrieved from https://www.nsif.mu/wp-content/uploads/2018/06/NCSR-Strat-plan-Final.pdf

National Empowerment Fund. (2011). Extract from budget speech. Retrieved from http://www.nef.mu/csr/budget_speech.php

Nyberg, D., Spicer, A., & Wright, C. (2013). Incorporating citizens: Corporate political engagement with climate change in Australia. *Organization*, *20*(3), 433–453.

Rambaree, K., Powers, M. C. F., & Smith, R. J. (2019). Ecosocial work and social change in community practice. *Journal of Community Practice*. Advance online publication. doi:10.1080/10705422.2019.1660516

Ross, D. (2013). Social work and the struggle for corporate social responsibility. In M. Gray, J. Coates, & T. Hetherington (Eds.), *Environmental social work* (pp. 193–210). London: Routledge.

Ross, J. A., & Zepeda, L. (2011). Wetland restoration, environmental justice and food security in the lower 9th ward. *Environmental Justice*, *4*(2), 101–108.

Schlosberg, D. (2013). Theorising environmental justice: The expanding sphere of a discourse. *Environmental Politics*, *22*(1), 37–55. doi:10.1080/09644016.2013.755387

Stein, R. (2004). *New perspectives on environmental justice: Gender, sexuality and activism*. New Brunswick, NJ: Rutgers University Press.

Svirydzenka, K., & Petri, M. (2017). Mauritius: The drivers of growth – can the past be extended? *Journal of Banking and Financial Economics*, *2*(8), 54–83. doi:10.7172/2353-6845.jbfe

Tandrayen-Ragoobur, V., & Kasseeah, H. (2018). Informal social protection: A case study of a small island economy. *International Journal of Public Policy*, *14*(5/6), 391–407.

# Crisis, constitutionalism and the geographies of belonging: Indian eco-politics in the Anthropocene

Ramya K. Tella

**ABSTRACT**

The Anthropocene – as a geological and cultural epoch – brings with it new modes of articulating environmental crisis and citizenship for peoples in India. This essay illustrates how space, scale and identity can be located within contemporary climate political imaginations through the normative canvas of the Indian Constitution. It argues that by thinking about climate change as a collection of dualisms – past/present, near/faraway, futures/futurelessness – Indian vocabularies of constitutionalism can inform authoritative understandings of crisis and citizenship. The article makes use of empirical material collected between 2016–17 for the author's doctoral research. It argues that Indian climate change elites characterize crisis in distinctive ways that draw attention to the symbolic and material implications of interrogating the shifting geographies of belonging in India. It shows how they make use of a framework of constitutionalism that is simultaneously attentive to the semantics of both planetary and subaltern thinking in the Anthropocene.

## Introduction

As a reference to a geological epoch, the Anthropocene is a term that denotes the pervasiveness of the impacts of humankind on their natural environment in recent centuries (Crutzen, 2006; Steffen, Crutzen, & McNeill, 2007) or, framed otherwise, it denotes the role of 'human agency' in shaping the geological cycles of the earth (Latour, 2017, p. 37). While the exact point of its emergence remains a source of continuing debate within the physical and social sciences, the Industrial Revolution remains a popular lens (Crutzen & Steffen, 2003; Smith & Zeder, 2013; Zalasiewicz et al., 2015) for 'Anthropocene origin stories' (Robin, 2013, p. 332). Through this lens, it is possible to observe how the diverse, and frequently divergent, imaginings of the Anthropocene become enmeshed in negotiations of global climate change politics and create a critical moment for constitutional epistemologies in the civic sphere (see Kotzé, 2019; Kotzé & Muzangaza, 2018, for a discussion on global environmental constitutionalism). These epistemologies as discursive frameworks, highlight the complexities and contestations that emerge around narrations of human-geophysical processes in languages of geographical and geopolitical singularity; the 'anthropos' serves as the core characterological figure of the Anthropocene and an active signifier of moral and political histories (Latour, 2017, p. 38).

This article is based on the premise that the current epoch is reflective of a critical constitutional moment. It argues that the longitudinal space–time politics of climate change calls forth new vocabularies of crisis and belonging that implicate histories of development in experiences of 'ongoingness' (Haraway, 2015, p. 160) and visions for the future, articulated in terms of futurelessness. These vocabularies are negotiated through a co-production of knowledge and society (Jasanoff, 2004) and point to the continuing significance of space to cultural narratives of environmental change (Hulme, 2008). It shows through empirical evidence that the Indian Constitution, as a normative standard for the formulation of natural-cultural identities, provides the raw materials for new imaginings of space, place and belonging.

This article makes use of the concept of civic epistemologies as initially formulated by scholars of science and technology studies (STS), which refers to the cultures of national knowledge making that inform the ways in which citizens understand and represent their environments (Jasanoff, 2010). This approach makes possible a comprehensive understanding of what it means to occupy a marginalized geographical sphere (see Fisher, 2015) – the global South – and to continuously produce and re-produce identity through the instrumentalization of a set of creative spatial–temporal dualisms. These include the past/present, the near/faraway and futures/futurelessness, which form the basis of situated eco-political discourses.[1] The fabric for the construction and mediation of these dualisms, and indeed the construction of the knowledges of climate change, is lent by intricately scripted narratives of constitutionalism, which in the process of demarcating what spaces and places ought to mean to publics, also simultaneously set out the contours of a politics of belonging; this politics is refracted through new and old nationalisms, and notions of environmental citizenship.

## Thinking law for the Anthropocene

The Anthropocene is reflective of a site of 'ontological-epistemological-ethical tension' (Nakagawa & Payne, 2019, p. 560) that is of planetary proportions. Importantly, the discursive constitution of such epochal change eludes a homogenous reading of the anthropos as a universal human citizenry in possession of an equal ability to impact on the geological foundations of the Earth (Grear, 2015). It intersects with scholarship in the legal sphere that speaks to framings that travel between the local and global spheres of sense making, to produce a new political rationality for the Anthropocene – through a 'modest (un)knowing' and a 'responsible (re)knowing' of environment and climate change (Nakagawa & Payne, 2019, p. 560); 'worlding' as a process of 'self-reflection and dialogue' (Inoue, 2018, p. 27) can be considered an example of one such framing. Nakagawa and Payne's phrase, 'Anthropocene knowledge ecology' (2019, p.560), and Santos (2016, p. 18) work on 'epistemicide' and the 'epistemologies of the South' can together show how collectives in the global South – following histories of colonial oppression, and cultural and epistemic erasure – make use of lenses that are drawn into a movement between locality and globality, to produce a new moral and political rationality for the Anthropocene, through a re-presentation of constitutionalism. Importantly, this resonates with scholarship from within the field of STS on the ethical drivers and implications of 'making global kinds of knowledge' (Hulme, 2010, p. 559).

An ethical and political interpretation of what it means to inhabit the Anthropocene is arguably enriched by the study of marginalized epistemologies of, and from, the global South (see Santos, 2016). This is also a point that is captured by Schmidt, Brown, and Orr (2016, pp. 191–192), who caution against an envisioning of climate change actions in the Anthropocene that subtly re-produce new configurations of the West's 'civilizing mission' through the operationalization of a green economy; Chaturvedi and Doyle (2015, p. 61) too, have shown that languages of greening rest on the establishment of a false relationship of equivalence between Southern development through industrialization and 'moral recklessness' in the present. These examples are illustrative of the modalities of global knowledge production (see Jazeel & McFarlane, 2010); they necessitate a comprehensive reading of the Anthropocene in terms of the numerous cultural ontologies that are in circulation – ontologies that do not map readily onto the epistemic rationalities of the global North (see Goldman, Turner, & Daly, 2018). It is at this confluence of conversations on culture, time, space and identity that the geographies of Anthropocene can be fruitfully revisited to evaluate ideas of personhood, space and rights, through the voices of individuals who are emplaced in simultaneous and overlapping *climatescapes* (Holzapfel, 2015) and *lawscapes* (Philippopoulos-Mihalopoulos, 2015).

Thinking law in the Anthropocene thus requires a shift from approaching the epoch as a 'temporal condition' to thinking about it as a differentiated 'cultural model' (Delanty & Mota, 2017, pp. 10–11); the key facets of this latter approach are temporality, subjectivity, knowledge and governance (Ibid., p.10). In India, they may intersect with imaginings of space through acts of boundary-making (Abraham, 2000) – these acts reflect the moral, cultural and political grammar of the postcolonial nation, its identities and norms of citizenship. The production of knowledge occurs concurrently with the production of a 'national inside' and a 'global outside' (Stripple, 2005, p. 148). As separate domains of interaction, these categories inform the conceptualization of climate justice as an aspirational ideal with scalar particularities (see Fisher, 2015; Tanner & Allouche, 2011); they encompasses inter-, as well as intra-generational norms of justice (Gonzalez, 2015). This understanding draws on a consciousness of the geographies of difference that circulate at the international level, owing to the historical consequences of colonialism, and subsequently of capitalism, and at the domestic level, where the quest for postcolonial modernity can, equally, conceal technocratic regimes of environmental governance (Baxi, 2001).

The Indian climatescape – understood as a spatial, temporal and cultural formation – is characterized by the prevalence of quite distinctive environmental rationalities that produce plural and frequently, contradictory, forms of knowledge. This idea of a climatescape (Holzapfel, 2015) maps onto the Idea of a 'lawscape,' as proposed by Philippopoulos-Mihalopoulos (2015, p. 86), to describe an 'ontological [spatial] category' that travels between past and present and plays out through dynamics of 'in/visibilisation.' In the lawscape, spatiality is itself a 'between thing,' located in the interstices between bodies, desires and disciplinary orientations (Ibid., p. 185). This article thus draws the various theoretical perspectives explored above into a discussion of law and climate change through the voices of Indian elites. While the majority of the perspectives from the literature on law in the Anthropocene tend to pertain to the sphere of international law and global constitutionalism, I show through empirical evidence the significance of contextualizing the languages of climate change against an evolving backdrop of Indian constitutionalism.

## Sources and methods

This article makes use of a method of narrative analysis and is based on 27 semi-structured interviews conducted with members of the Indian climate change elite between 2016–2018. These actors spanned the spheres of academia, bureaucracy, policy, politics, science, and journalism and were approached following methods of purposive and snowball sampling.[2] The selection of these actors was based on a criterion of individual visibility in a national context, where the naming of respondents as actors emerged from a consideration of the material consequences of climate political theater in India (see Death, 2011, for a discussion on theatricality in climate change politics). It was operationalized through searches carried out on Google and on the website of a popular Indian research journal called the Economic and Political Weekly (EPW). In the case of the former, the search terms 'Indian climate science' helped in the identification of scientific actors who were based in prominent research institutions such as the Indian Institutes of Technology (IITs), the Indian Institute of Science (IISc) and Ministries of the Central Government. The search term 'climate' on the website of the EPW, revealed a host of other actors from within the remaining spheres who have been active in formulating Indian climate change positions on the national level; they consisted of actors from The Energy and Resources Institute (TERI) and the Centre for Policy Research (CPR), among others. Additionally, meetings with a high-level political actor and a former diplomat also opened up access to a number of other diplomatic and scientific actors who are closely involved with the construction of climate change knowledges and the drafting of national policies. This article explores the usage of constitutionalism by these actors as a principle that underwrites narratives on 'Indian-ness' (Billett, 2010, p. 13) and the framing of climate change as a crisis. It makes use of a method of textual analysis, developed through the preliminary identification of key narrative terminologies or codes from the interview data, which refer to constitutional principles.

## Constitutionalism in Indian climate politics

The politics of climate change in India is best described in terms of its duality – as a country with widespread poverty and low per capita emissions, and; as a large economy whose current developmental trajectory is characterized by increasing carbon emissions (Dubash, 2013). Principles of equity and historical responsibility have guided Indian contributions to global discourses of climate change through an emphasis on the per capita principle (Thaker & Leiserowitz, 2014), distinctions between 'survival emissions' and 'luxury emissions' (Agarwal & Narain, 2003, p. 3), 'environmental colonialism' (Ibid., p.1) and the binary of global North and South (Joshi, 2014). Indian narratives from across diverse disciplines display a strong belief in the science of climate change. However, these narratives also interact with contestations over the sources and legitimacy of scientific knowledge (Biermann, 2001; Kandlikar & Sagar, 1999); one prominent example of this is Himalayagate – a controversy that emerged from the inclusion of an incorrect prediction in the Fourth Assessment Report of the IPCC, which stated that the Himalayan glaciers would disappear by the year 2035 (Carrington, 2010; Mahony, 2014). It prompted the formation of the Indian Network for Climate Change Asssessment (INCCA), whose legitimacy rested on its territorial emplacement within the nation, rather than in western spaces of knowledge production, which the IPCC symbolized (Mahony, 2014).

More recently, climate change narratives in India have had to contend with growing expressions of religious nationalism and associated attempts to restructure the language of environmental engagement across the country – an important example of this is the unofficial re-naming of the Ministry of Earth Sciences as the *Bharat Mata Mantralaya* (the Ministry of Mother India) (Mohan, 2019), which signifies an assertion of a Hinduized pursuit of scientific knowledge at the national scale through a gendered rendition of the nation in the female form (see Ramaswamy, 2002). This particular instance, and the wider phenomenon that it signifies jointly point to a crisis of the constitutionally ingrained principle of secularism – a principle that informs understandings of democracy and scientific temper in India (see Arnold, 2013, on Nehruvian science; see also, Subramaniam, 2000 and 2019, on religion and scientific knowledge in India).[3] Together, the various aspects discussed above, shape national-level understandings of climate change, and prompt the tracing of linkages to constitutional framings of citizenship and rights in India.

The Constitution forms a critical component of Indian narratives on climate change and informs the 'moral meteorology' (Burman, 2017, p. 926) of actors.[4] The knowledges that actors produce and perform, attain resonance and visibility through references to the constitutional text, and its wider symbolism as a pre-eminent framework for a culturally embedded imagining of norms. The Constitution informs Indian Anthropocene imaginaries in ways that allow actors to construct understandings of 'things as they ought to be' (Jasanoff, 2011, p. 634), and forms a basis for the interpretation of the civic epistemologies of climate change in India and their connections to the contemplation of crisis. The framing of climate change as a 'perfect moral storm' (Gardiner, 2006, p. 398) when refracted through this case study of India, reveals heterogeneous imaginings of space, place and ideas of belonging. This is aided by the work of Adger, Butler, and Walker-Springett (2017), who observe that morality in discourses of climate change weave through ideas that can be organized along two strands, namely, vulnerability-, and system-based accounts. In the former, notions of 'solidarity,' 'entitlements,' 'fairness in burdens' and 'protection from harm' assume a significant role while the latter consist of 'respect for authority,' 'stability and system preservation,' 'duty and responsibility' founded on understandings of 'citizenship' and 'patriotism,' and 'sanctity, purity, and naturalness' (Ibid., p. 376). The actors interviewed in this case study reflected these ideas in different ways. They showed that the framing of crisis when read through the language of the Indian Constitution is open to creative modes of interpretation.

For example, one actor from a prominent foreign policy research organization in India articulated a narrative of constitutional authority by constructing a linkage between legal and religious languages in India. He stated:

> The Indian narrative is compelling because it is constitutionally ingrained in Article 21. [It is] civilisationally deep-rooted – ancient texts like the Vedas, the principles of Hinduism or even Buddhism are culturally constructed. (P1 Interview)[5]

This narrative cuts across a diverse set of themes which are as follows: the right to life and protection from harm, as indicated by the actor's reference to Article 21 of the Indian

Constitution; respect for authority through a reference to higher legal doctrine and; sanctity, purity, and naturalness through a culturally constructed religious-civilisational narrative that eschews overconsumption – the actor in this particular instance cited the practice of 'Yoga' as an Indian response to 'western greed and overspending' (P1 Interview).[6] While this narrative constitutes one means of constructing an Indian vision for the Anthropocene through an interweaving of religious and constitutional geographies, there exist still other narratives that locate the significance of the Indian Constitution in a more explicit articulation of the identity of the newly independent, postcolonial nation. Thus, addressing the historical significance of a constitutional narrative of climate change, a former diplomat stated:

> Indian foreign policy was shaped by Jawaharlal Nehru and the values of India's independence movement – these were very much imbibed. Actually, there's an Article in the [Indian] Constitution which makes it obligatory for India to play a role in the establishment of peace and cooperation in the world. (D1 Interview)[7]

The Constitutional Article 51, which forms a part of the Directive Principles of State Policy is concerned with the promotion of international peace and security. It contains sub-clauses that each pertain to the maintenance of international relationships, the cultivation of respect for international law and treaty commitments, and the encouragement of international political settlement through intervention as an arbiter. Together, these share a direct link with the formulation of the Non-Aligned Movement (NAM) as a movement that was designed to function as an alternative to Cold War polarities, alongside an adherence to principles of sovereignty, and the preservation and protection of national interest (see Abraham, 2008).

The invocation of the NAM in a climate change context cuts across a number of ideas which help to frame an Indian articulation of morality that is constitutionally rooted in Fundamental Rights and Duties. These are as follows: solidarity with other developmentally similar nations; protection from harm as a possible signification of the dangers posed by climate change as well as those that arise from the ontology of the climate change question and its location in the industrialization of the west; respect for authority, which can be viewed in three intersecting ways as respect for Nehru – a political figure, respect for the Indian Constitution – a source of higher authority, respect for international law – as an authoritative code of global conduct; stability and system preservation, which pertain to the preservation of the natural environment – both nationally (Article 48A) by protecting forests and wildlife, and internationally (Article 51) through the settlement of boundary disputes, arising over land (territory) or bodies (migrants, refugees), and the preservation of a global environmental system within a cosmopolitan conception of the environment in the Anthropocene that recognizes the differentiated entitlements of specific clusters of nations. Importantly, the contributions of the Constitution to the formulation of a cultural sub-structure for Indian climate change imaginations is also reflected in other narratives which vocalize constitutional precepts as self-legitimating dimensions of national storylines:

> Whenever you take India's position about any significant matter, it would start off from the Constitution of India. As you are aware the Indian Constitution is egalitarian … so whether it is climate or trade or security, so the first sort of anchor is the Constitution, and then the relevant legislation, and then naturally the governmental power and its philosophy and all

> of that. The importance the Indian position accords to poverty eradication ... that's the economic need of the country, but it's also a constitutional imperative. Similarly, the importance accorded to equity, because equity in climate change parlance is the rights of each human being whether rich or poor. Now, India is a big country for equity, so that also arises from overall domestic imperatives [of poverty eradication]. (D2 Interview)[8]

Diplomat D2's narrative produces a constitutionally formulated idea of India that shares some important parallels with the narrative discussed immediately prior – of D1. However, unlike the earlier narrative, it is more specifically focused on amplifying India's identity as a developing nation. The actor sets out a hierarchical conception of cultural and moral authority, and locates the constitutional text as the fundamental basis for Indian civic epistemologies, followed by the more variable dimension of national government and political philosophy. The justification underlying this design of a climate change narrative by the actor is rooted in memories of colonialism and an active archaeology of the present, which is presented as a consequence of capitalism.

The actor therefore identifies equity and the rights of each human being (Indian citizens) as essential to a national story of crisis that adopts the per capita basis for emissions entitlements as a source of authority and reifies the narrative focus by incorporating principles of fairness in burdens and protection from harm, through a reference to the objective of poverty eradication. He invokes a respect for authority through an emphasis on the legal power of the Constitution and presents an idea of duty and responsibility that responds to the particularities of the moral-legal substructure of an Indian climate change narrative; he describes this in terms of a commitment to the Indian citizenry through the discursive enforcement of the responsibility of the global North to the South. However, the domestic dissonances in this actor's approach to climate change as a crisis are worth illustrating in their entirety. In particular, the actor's subsequent framing of 'climate change as war' and 'development through destruction' (D2 Interview) produce a domestic narrative that contrasts starkly with the actor's approach to the international politics of climate change. It reveals how constructions of the phenomenon are indicative of moral crises:

> Technological breakthroughs come during – not as a principle, but generally – periods of stress. So like [take the example of] wars – a number of technological breakthroughs have come. And what we have to do for climate change is also some kind of a war. (D2 Interview)

(and)

> We need to develop, we need to have factories, we need to have dams, we need to have destruc ... not destruction, but we need to sort of remove the trees and all for development. (D2 Interview)

The specific languages that state actors use to represent climate change to domestic audiences reveal that marginalized bodies and spaces within the national climatescape are frequently rendered invisible in portrayals of morality, rights and justice through a substitution of politics with a discourse of development from above. It presents a paradox when viewed in relation to the constitutional emphasis on the safeguarding of the rights of minorities through the right to equality before the law (Article 14), the blanket 'prohibition of discrimination' against minorities (Article 15(1)) and the establishment of the rights of all individuals to public spaces (Article 15 (2)). This is further supported by the Fundamental Duties, which emphasize the promotion of 'special care' (Article 46) in the protection of the socio-economic interests of marginalized sections of the Indian population – the Scheduled Castes (SCs) and Scheduled Tribes (STs) – who continue to be subjected to institutionalized discriminatory practices. However, when contrasted to diplomat D2's narrative, there are numerous sources of tension that are revealed. First, while D2's endorsement of a process of economic development that is contingent on technological progress forms a key attribute of postcolonial modernity (see Arnold, 2013), the actor's approach is antithetical to the languages of peace and solidarity that the Constitution seeks to promote. Second, the diplomat fails to acknowledge the discriminatory nature of development that is proposed, given the enormity of recent evidence from the domestic politics of environment in India that highlights the institutional oversight of historical claims to space on sub-national scales (see Yenneti, Day, & Golubchikov, 2016). Further, as this research shows, actors also employ constitutional languages as a means of lending credibility to a broader narrative of scientific paternalism, which assumes varying versions of the following statement:

> Science [is] to be used for the society, by the society, of the society. It's not only democracy [that is meant for people], it is also the science and technology to be utilized by the people, for the people, of the people. Then only country will progress. Right? (S1 Interview)[9]

This establishment of a story of scientific progress, as an idea that exists in the service of the Indian peoples, is equated to the very essence of democracy by actors and constitutes an extension of the script of the Preamble of the Indian Constitution, which ventriloquizes the imagined voice of *the peoples of India*. However, this perspective on the ability of science to guide progress in equal and neutral ways, is also critiqued by non-state actors who draw attention to the unequal geographies of technoscientific progress. This may be evidenced by the prevalence of discourses around forest rights, nuclear energy, large dams and genetic modification in India, which draw attention to an out-maneuvering of democratic environmentalism and dissent by the modern state (A1 Interview, J1 Interview[10]). These different perspectives illustrate that even as national climate change narratives attempt to co-opt constitutional principles as a means of enhancing their authenticity or 'Indian-ness' (Billett, 2010, p. 13), the social, political and cultural dimensions of contemplating climate as crisis, are open to plural modes of interpretation.

## 'Spacing' crisis in Anthropocene geographies

The articulation of climate change as a crisis occurs in a number of different ways. One important lens that actors employ is the dualism of past and present – as a mutually reinforcing dimension of Indian civic epistemologies of climate change. The majority of its applications pertain to the international sphere of climate change politics and are viewed by actors as a necessary prelude to any engagement with the politics of the future citizen. In an incisive way, this temporal lens is enmeshed in a politics of space that locates the origins of its historical trajectory in the advent of colonialism and the Industrial Revolution, and the subsequent proliferation of uneven development geographies. However, the various perspectives from this study also depict the challenges faced by actors in navigating imaginings of the Anthropocene in the global South –

through encounters with the dualism of the near and the faraway. This is captured in the following statement by a policy actor:

> At a very large level, there's this whole sort of existential question that the idea of climate change is kind of a driver of Anthropocene, a whole new epoch where humans are able to shape the biosphere, and you know, catastrophic climate change. That's so hard to grapple with in your head. So, the things that tend to influence Indians are more narrow. (P2 Interview)[11]

The above statement is illustrative of a conception of climate change that points to the material and imaginative disjunctures that are immanent in any proposal to produce a totalizing global description of climate change and the Anthropocene. It challenges the idea of a 'view from nowhere' (see Shapin, 1998, p. 5), by deliberately locating the politics of climate change within an Indian imagining of crisis – as something that is in the *here* and *now*, rather than approaching it as an unmoored object that ought to be tackled for the benefit of a future global citizenry. The actor's emphasis on the recognition of a proximate lexicon of climate change, is based on the wider significance of the fulfillment of basic needs in India.[12]

Ideas of poverty alleviation, equity and human rights, have informed Indian approaches to climate change by undergirding perceptions of crisis; these have contributed to understandings of temporal politics and human agency by forcing a differentiated narration of the geographies of climate responsibility. They inform the way knowledges are produced through the employment of vocabularies that pertain to a re-assertion of nation state politics alongside an appreciation of the need for a cosmopolitan approach to climate change (see Beck, 2015). The nation state, as a component of Indian imaginings of crisis, is organically folded into an understanding of time – of the impact of human agency on geological time, through the assertion of the postcolonial identities of the Indian citizenry. Here, the project of decolonizing (Radcliffe, 2017) the discourse of climate crisis can be enriched by a systematic engagement with the Indian Constitution, which offers a means of situating climate futures in understandings of marginalized presents. It makes possible a cultural and affectual refocusing of vulnerable geographical scales within global-level discourses of a singular planetary future:

> How can we develop earth system justice and how can we develop multi-species solidarity? Science, which began in wonder, is now ending in destruction and we have lost the capacity to be surprised or astonished. And in an interesting way, has climate change made a difference? It is a change of its own type. It is not a conventional concern. Environmental change and climate change can only be dealt with on a global level. But futurelessness is a condition and it becomes universalized. (A2 Interview)[13]

Ideas of crisis, as the majority of Indian perspectives on climate change in this study illustrate, are characterized by understandings of history, politics and morality. This is observable in both explicit and implicit references to colonialism, and also, in the divergences that open up among elite perspectives in their treatment of the language of crisis. These modes of engagement are important to map out in the wider context of viewing postcoloniality as a fundamentally plural condition, particularly as it concerns environmental politics and climate change in India:

> See, if you're in a boat, if you have a life jacket you jump into the sea. If you don't have a life jacket, who is going to save you? So development is a life jacket. (P3 Interview)[14]

The metaphor of the lifejacket serves as a specific reaction to the characterization of climate change as a post-political concern; this shares direct linkages to the constitutional rights to equality (Article 14), the prevention of discrimination (Article 15) and life and personal liberty (Article 21). It seeks to produce a postcolonial perspective on climate change that holds a differentiated frame on the figure of the anthropos to show how the discourse of crisis, as represented in the statement above through a reference to a figurative sea, can enter narratives. However, while a number of narratives in this study make use of a historical memory of colonialism as a means of explicating what is meant by responsibility, they evolve quite differently among this cohort of Indian actors to reveal varying expressions of crisis and environmental citizenship.

The imaginative negotiation of crisis is constitutive of a site at which the epistemic geographies of climate change come to play an important role, especially in the narratives constructed by state actors (diplomats and scientists). Driven by a belief that the knowledge produced by the IPCC (Intergovernmental Panel on Climate Change) is inaccurate, these actors seek to produce a climate change narrative that rests on the negation of 'Northern knowledge' through a re-assertion of the role of the nation state as *the* authoritative epistemic site. Indian climate change knowledges, understood as knowledges produced within India and by Indian scientists, is thus accorded the highest importance in these narratives:

> Actual manifestations of climate change in India as monitored by India's scientific establishment do not yet conform to the projections of the IPCC. It is difficult to establish (or argue) that the weather related adverse events that have occurred in the past decade (e.g. extreme rainfall events in Mumbai, Srinagar, Chennai) are way out of line with historical experience. Nor have the promised floods in the Indo-Gangetic plain from melting glaciers happened, and neither has the projected unmanageable tide of climate refugees due to sea-level rise from Bangladesh. Changes in monsoon patterns that have so far been observed do not quite conform to the IPCC story, are not particularly severe, nor out of line with historical cycles. (D3 Interview)[15]

The above and other similar perspectives from this research are partially a reaction to the IPCC's erroneous prediction of Himalayan glacial melt in 2010 (see Mahony, 2014, as already discussed). Additionally, these also symbolize the placement of certain Indian ideas about knowledge in understandings of sovereignty – in this case, through epistemic sovereignty, territorial integrity, and an emphasis on the Indian scientific establishment. These are used by actors such as diplomat D3 and D2 to oppose global narratives of fear through recourse to perceptions of sound national science which are frequently communicated through a commitment to nationalism. Domestically, these constructions of nationalism for state actors are reified through their encounters with specific climate political characters such as refugees.

While the linkages between migration (Stojanov, Boas, Kelman, & Duží, 2017) or conflict (Selby, Dahi, Fröhlich, & Hulme, 2017), for instance, and climate change, resist attributions of causality, references to contemporary spatial processes and forms of public protest, particularly as employed by non-state actors, rely on constitutional languages of equality and protection. These are used to construct framings of rights to survival that are contingent on the recognition of the socio-cultural, political and ecological vulnerabilities of marginalized communities. For example, the recent abrogation of Article 370 of the Indian Constitution (Kumar, 2019), which had previously granted special or autonomous status to the

Indian state of Jammu and Kashmir, and the proposals for the implementation of the National Register of Citizens (NRC) in Assam and the introduction of the Citizenship Amendment Bill (CAB) (Poddar, 2018; Sharma, 2019), are reflective of the unconstitutional actions of institutions of state. These actions have the potential to affect the ability of persons to adapt to climatic changes.

Importantly, the climate narratives produced by civil society actors stand in stark contrast to the narratives of state. The following perspective on crisis offers a useful slant:

> If you see what's happening in Dhaka for instance, climate refugees. It's a reality in today's world. 2000 people come to Dhaka every day. Why? Because they can't live in those estuarine lands they were growing. It's all flooded. (J1 Interview)

The appeal of this narrative on climate refugees is significant to an understanding of the geographies of crisis in India. It also indicates, more broadly, that there exist differences in the way crisis is articulated by state and non-state actors, which become evident in a comparison of the two statements shown above (J1 and D3, Interviews). Crucially, the constitutional conceptualization of citizenship serves as both a source of delimitation and emancipation; on the one hand, it articulates citizenship in terms of birth, ancestry and residence; on the other, and importantly, it speaks to an ethics of care that is premised on the right to life. The latter may be extended to the protection of migrant populations, within the ambit of human rights and security. Here, the constitutional protections offered to domestic migrants draw on a different set of constitutional articles, which consist of the right to freedom of movement, residence and settlement (Article 19d and 19e); these provisions make it possible to recognize the legitimacy of both national and international geographical mobilities that may be influenced by a complex combination of socio-economic and ecological factors. This could apply to migratory narratives such as the following:

> The person who is living in the dry areas of Rayalaseema [a region in the Indian state of Andhra Pradesh] or on the coast of Vishakapatnam [a town in Southern India], he has no option. He is there, and he is already vulnerable. He is poor, and he's living on the coast or on the banks of the Ganga and when floods come, his house is going to get sunk. The other guy has migrated out. So people who are already vulnerable will be more vulnerable to climate change. (P4 Interview)[16]

As these perspectives show, 'the near' and 'the faraway' (Krishna, 2013, p. 125) are frames that are not only relevant to Indian engagement with the international politics of climate change but also symbolize the dissonances that inhere in domestic sites of imagination. This was also observable in a distinction that one actor made between 'India' as a reference to the vision of the state, and 'the people of India' as a reference to marginalized citizens and sub-national geographies (J2 Interview).[17] Together, these perspectives point to the possibility of approaching *Indian-ness* in elite imaginings of crisis as multiple – it is a cultural, political and moral construct that interacts with ideas about the organization of national space to produce an intimate politics of climate change that is challenged by de-politicizing currents at both the global and domestic scales.

In this research, the construction of a nationalistic identity by actors of state converges with articulations of postcolonial modernity; the authority of these narratives rests on a simultaneous negation of *other* sub-national voices that are exterior to the machineries of state. Two examples of the other are 'activists' and 'refugees,' who rupture institutional imaginaries of the good environmental citizen. This citizen, as the state narratives in this study show, may be understood as an individual who subscribes to the practices of technoscientific modernity, through the production of a politically and ideologically homogenous national inside space that is separate and impermeable to the global outside space. Strikingly, the constitutional rights of citizens to freedom of speech and expression, and public assembly, are affected by the discursive suppression of dissent within this inside space that is enunciated through labels that question the nationalism of certain individuals and groups. For example, one diplomat made the following statement:

> India's approach to climate change has episodically been raised in the media by the environmental activist community, but in recent years have been largely discounted by the media, possibly because they are now seen as either "useful fools" or otherwise pursuing extranational agendas. (D3 Interview)

For the non-state actors who seek to critique this view in their own ways, the challenge is not only one of establishing the credibility of their respective climate change knowledges, but also of countering a new language of terror that shifts the focus away from the phenomenon of climate change to the identification of individuals who express dissent. The tug between the opposing frames of the good and bad environmental citizen may even play out, as one respondent observed, through the language of 'green terrorism' that is used to silence disagreements within the domestic landscape of environmental politics (J1 Interview).

The vocabulary of crisis as this case study of elite Indian actors shows, encompasses binaries of nationalism and extra-nationalism that are used to form value judgments about certain groups of individuals who violate the state's code of the good environmental citizen. These examples are also indicative of the broader tensions inherent in thinking about the lawscape and climatescape as a continuous imaginative canvas, where the tenets of the Indian Constitution, particularly as they concern the provision of Fundamental Rights to individuals are obfuscated from above in the context of dissenting bodies.

## Conclusion

This article has illustrated how the contemplation of crisis and the geographies of belonging in the climatescape are complicated in fruitful ways by the emergence of Anthropocene thinking. Indian knowledges of climate change, as shown here, both produce and resist paradoxical constructions of modernity, by engaging ideas of development, equality and sovereignty. These are accompanied by a continuous movement across scales in elite narratives, which points to the moral and ethical implications of articulating situated stories of crisis. Importantly, the usage of constitutional languages is, in itself, quite varied across the Indian climatescape.

An intimate reading of constitutionalism in this context thus makes it possible for actors to travel between the local and global scales of politics, while contending with the framing of climate change as a 'de-spatialising mythology' (Philippopoulos-Mihalopoulos, 2015, p. 30). The climate actor, when framed as 'the person who *comes* to the law' (Philippopoulos-Mihalopoulos, 2015, p. 185) [italics in original], can immediately be located within an emplaced story of the Anthropocene that recognizes the nested hierarchies (Grear, 2015) that determine how climate change is ultimately represented at different scales. This is captured in the following statement by Philippopoulos-Mihalopoulos (2015,

p. 185): 'The movement [to the law] is specific, spatially inscribed, bodily demanding, and expresses a "here" that deserves to be heard at least as loudly as "now".'

The narratives discussed here have attempted to show that there is much to gain from an open-ended reading of the Indian Constitution that expands the applicability of rights to include an evolving understanding of environmental citizenship. This is made possible by thinking through crisis in terms of a set of dualisms – past/present, near/faraway, futures/futurelessness – which are operationalized simultaneously in elite imaginings, and are both dismantled and reinforced by different actors in this study.

## Notes

1. The usage of the term eco-politics draws on an understanding of the close linkages between cultural constructions of environment and contestations around space. It draws on a critical examination of environment and sovereignty by Litfin (1997) that points to the significance of exploring evolving systems of legitimacy across scales, through a focus on state and non-state actors.
2. In citations that pertain to interview data, the prefix 'A' refers to academics, 'D' to diplomats, 'J' to journalists, 'P' to policy actors and 'S' to scientists.
3. Subramaniam (2000, pp. 73–74; 2019, p. 69) engages in a comprehensive examination of the interactions between ideas of Hindu nationalism, science and democracy in India, which are characterised by the active acceptance of capitalism, and western science and technology, alongside a glorification of ancient Hinduism; together, these comprise a form of 'archaic modernity'.
4. Burman (2017, p. 922) discusses this concept in relation to 'global cognitive justice' and does not engage with constitutionalism. It is introduced in this article owing to its significance to the framing of an idea of constitutionalism in the global South.
5. Author's interview with a policy expert from the Institute of Defence Studies and Analyses (IDSA), 13th February, 2017.
6. In recent years especially, Yoga has occupied an important position in imaginings of climate change and the good environmental citizen, where it has been deployed as an instrument of 'bionationalism' (Subramaniam, 2019, pp. 10–11) — the proponents of this discourse in Government, for instance, have drawn on individualised representations of environmental responsibility that are premised on Hindu constructions of physical discipline and values. This has been reinforced by key institutions of state, such as the Ministry of Environment, Forests and Climate Change and the Ministry of Health and Family Welfare, as too by actors such as Narendra Modi.
7. Author's interview with a former diplomat, 29th January, 2017.
8. Author's interview with a former diplomat and climate change negotiator, 13th September, 2016.
9. Author's interview with a scientist in a Ministry of the Central Government, 15th February, 2017.
10. Author's interview with an with an academic, 21st August, 2017, and with an independent journalist, 22nd November, 2016.
11. Author's interview with a policy expert from the Centre for Policy Research (CPR), New Delhi, 28th August, 2016.
12. The Constitutional articles pertaining to the right to equality (Article 14) and the right to life (Article, 21) are important in this regard.
13. Author's interview with an academic, 6th May, 2017.
14. Author's interview with an independent policy expert, 18th February, 2017.
15. Author's interview with a former diplomat and climate change negotiator, 1st March, 2017.
16. Author's interview with a policy expert from the Delhi Science Forum (DSF), New Delhi, 25th August, 2016.

17. Author's interview with an independent journalist, 16th February, 2017.

## Acknowledgements

I thank my doctoral supervisors, George Adamson and Mike Hulme, for their encouragement and support. I also thank Benito Cao and Adela Alfonsi for their invaluable editorial guidance, and gratefully acknowledge the constructive suggestions given to me by the two anonymous reviewers.

## Disclosure statement

No potential conflict of interest was reported by the author.

## ORCID

*Ramya K. Tella* http://orcid.org/0000-0002-2902-9603

## References

Abraham, I. (2000). Landscape and postcolonial science. *Contributions to Indian Sociology, 34*(2), 163–187.

Abraham, I. (2008). From Bandung to NAM: Non-alignment and Indian foreign policy, 1947–65. *Commonwealth & Comparative Politics, 46*(2), 195–219.

Adger, W. N., Butler, C., & Walker-Springett, K. (2017). Moral reasoning in adaptation to climate change. *Environmental Politics, 26*(3), 371–390.

Agarwal, A., & Narain, S. ([1991] 2003). *Global warming in an unequal world: A case of environmental colonialism*. New Delhi: Centre for Science and Environment.

Arnold, D. (2013). Nehruvian science and postcolonial India. *Isis, 104*(2), 360–370.

Baxi, U. (2001). What happens next is up to you: Human rights at risks in dams and development. *American University International Law Review, 16*(6), 1507–1530.

Beck, U. (2015). Emancipatory catastrophism: What does it mean to climate change and risk society? *Current Sociology, 63*(1), 75–88.

Biermann, F. (2001). Big science, small impacts—in the South? The influence of global environmental assessments on expert communities in India. *Global Environmental Change, 11*(4), 297–309.

Billett, S. (2010). Dividing climate change: Global warming in the Indian mass media. *Climatic Change, 99*(1-2), 1–16.

Burman, A. (2017). The political ontology of climate change: Moral meteorology, climate justice, and the coloniality of reality in the Bolivian Andes. *Journal of Political Ecology, 24*(1), 921–930.

Carrington, D. (2010). IPCC officials admit mistake over melting Himalayan glaciers. *The Guardian*. Retrieved from https://www.theguardian.com/environment/2010/jan/20/ipcc-himalayan-glaciers-mistake

Chaturvedi, S., & Doyle, T. (2015). *Climate terror: A critical geopolitics of climate change*. Basingstoke: Palgrave Macmillan.

Crutzen, P. J. (2006). The "anthropocene". In E. Ehlers, & T. Krafft (Eds.), *Earth system science in the Anthropocene* (pp. 13–18). Berlin, Heidelberg: Springer.

Crutzen, P. J., & Steffen, W. (2003). How long have we been in the Anthropocene era? *Climatic Change, 61*(3), 251–257.
Death, C. (2011). Summit theatre: Exemplary governmentality and environmental diplomacy in Johannesburg and Copenhagen. *Environmental Politics, 20*(1), 1–19.
Delanty, G., & Mota, A. (2017). Governing the Anthropocene: Agency, governance, knowledge. *European Journal of Social Theory, 20*(1), 9–38.
Dubash, N. K. (2013). The politics of climate change in India: Narratives of equity and cobenefits. *Wiley Interdisciplinary Reviews: Climate Change, 4*(3), 191–201.
Fisher, S. (2015). The emerging geographies of climate justice. *The Geographical Journal, 181*(1), 73–82.
Gardiner, S. M. (2006). A perfect moral storm: Climate change, intergenerational ethics and the problem of moral corruption. *Environmental Values, 15*(3), 397–413.
Goldman, M. J., Turner, M. D., & Daly, M. (2018). A critical political ecology of human dimensions of climate change: Epistemology, ontology, and ethics. *Wiley Interdisciplinary Reviews: Climate Change, 9*(4), 1–13.
Gonzalez, C. G. (2015). Bridging the north-south divide: International environmental law in the anthropocene. *Pace Environmental Law Review, 32*(2), 407–434.
Grear, A. (2015). Deconstructing anthropos: A critical legal reflection on 'anthropocentric'law and anthropocene 'humanity'. *Law and Critique, 26*(3), 225–249.
Haraway, D. (2015). Anthropocene, capitalocene, plantationocene, chthulucene: Making kin. *Environmental Humanities, 6*(1), 159–165.
Holzapfel, A. S. (2015). From landscape to climatescape in contemporary Dance-Theater. In S. George-Graves (Ed.), *The Oxford Handbook of Dance and Theater* (pp. 401–428). New York: Oxford University Press.
Hulme, M. (2008). Geographical work at the boundaries of climate change. *Transactions of the Institute of British Geographers, 33*(1), 5–11.
Hulme, M. (2010). Problems with making and governing global kinds of knowledge. *Global Environmental Change, 20*(4), 558–564.
Inoue, C. Y. A. (2018). Worlding the study of global environmental politics in the Anthropocene: indigenous voices from the Amazon. *Global Environmental Politics, 18*(4), 25–42.
Jasanoff, S. (2004). The idiom of co-production. In S. Jasanoff (Ed.), *States of knowledge: The co-production of science and social order* (pp. 1–12). New York & London: Routledge.
Jasanoff, S. (2010). A new climate for society. *Theory, Culture & Society, 27*(2-3), 233–253.
Jasanoff, S. (2011). Constitutional moments in governing science and technology. *Science and Engineering Ethics, 17*(4), 621–638.
Jazeel, T., & McFarlane, C. (2010). The limits of responsibility: A postcolonial politics of academic knowledge production. *Transactions of the Institute of British Geographers, 35*(1), 109–124.
Joshi, S. (2014). Environmental justice discourses in Indian climate politics. *GeoJournal, 79*(6), 677–691.
Kandlikar, M., & Sagar, A. (1999). Climate change research and analysis in India: An integrated assessment of a south–north divide. *Global Environmental Change, 9*(2), 119–138.
Kotzé, L. (2019). A global environmental constitution for the Anthropocene? *Transnational Environmental Law, 8*(1), 11–33.
Kotzé, L. J., & Muzangaza, W. (2018). Constitutional international environmental law for the Anthropocene? *Review of European, Comparative & International Environmental Law, 27*(3), 278–292.
Krishna, S. (2013). IR and the postcolonial Novel: Nation and subjectivity in India. In S. Seth (Ed.), *Postcolonial Theory and international Relations* (pp. 124–143). Oxon & New York: Routledge.
Kumar, A. (2019). The constitutional legitimacy of abrogating article 370. *Economic & Political Weekly, 54*(38), 15.
Latour, B. (2017). Anthropology at the time of the Anthropocene: A personal view of what is to be studied. In M. Brightman, & J. Lewis (Eds.), *The anthropology of sustainability* (pp. 35–49). New York: Palgrave Macmillan.
Litfin, K. T. (1997). Sovereignty in world ecopolitics. *Mershon International Studies Review, 41*, 167–204.

Mahony, M. (2014). The predictive state: Science, territory and the future of the Indian climate. *Social Studies of Science*, *44*(1), 109–133.

Mohan, V. (2019). Earth Sciences can be Bharat Mata Mantralaya: Harsh Vardhan. *The Times of India*. Available at: https://timesofindia.indiatimes.com/india/earth-sciences-can-be-bharat-mata-mantralaya-harsh-vardhan/articleshow/67548920.cms. [Accessed on 1st October, 2019]

Nakagawa, Y., & Payne, P. G. (2019). Postcritical knowledge ecology in the Anthropocene. *Educational Philosophy and Theory*, *51*(6), 559–571.

Philippopoulos-Mihalopoulos, A. (2015). *Spatial justice: Body, lawscape, atmosphere*. Oxon & New York: Routledge.

Poddar, M. (2018). The citizenship (Amendment) bill, 2016: International law on religion-based discrimination and naturalisation law. *Indian Law Review*, *2*(1), 108–118.

Radcliffe, S. A. (2017). Decolonising geographical knowledges. *Transactions of the Institute of British Geographers*, *42*(3), 329–333.

Ramaswamy, S. (2002). Visualising India's geo-body: Globes, maps, bodyscapes. *Contributions to Indian Sociology*, *36*(1-2), 151–189.

Robin, L. (2013). Histories for changing times: Entering the Anthropocene? *Australian Historical Studies*, *44*(3), 329–340.

Santos, B. D. S. (2016). Epistemologies of the South and the future. *From the European South: a Transdisciplinary Journal of Postcolonial Humanities*, *1*, 17–29.

Schmidt, J. J., Brown, P. G., & Orr, C. J. (2016). Ethics in the Anthropocene: A research agenda. *The Anthropocene Review*, *3*(3), 188–200.

Selby, J., Dahi, O. S., Fröhlich, C., & Hulme, M. (2017). Climate change and the Syrian civil war revisited. *Political Geography*, *60*, 232–244.

Shapin, S. (1998). Placing the view from nowhere: Historical and sociological problems in the location of science. *Transactions of the Institute of British Geographers*, *23*(1), 5–12.

Sharma, C. (2019). Citizenship Amendment Bill 2016: Continuities and contestations with special reference to politics in Assam, India. *Asian Ethnicity*, *20*(4), 522–540.

Smith, B. D., & Zeder, M. A. (2013). The onset of the Anthropocene. *Anthropocene*, *4*, 8–13.

Steffen, W., Crutzen, P. J., & McNeill, J. R. (2007). The Anthropocene: Are humans now overwhelming the great forces of nature. *AMBIO: A Journal of the Human Environment*, *36*(8), 614–621.

Stojanov, R., Boas, I., Kelman, I., & Duží, B. (2017). Local expert experiences and perceptions of environmentally induced migration from Bangladesh to India. *Asia Pacific Viewpoint*, *58*(3), 347–361.

Stripple, J. (2005). Conclusions. In *Climate change after the international: Rethinking security, territory and authority* (pp. 137–155). Lund: Lund University.

Subramaniam, B. (2000). Archaic modernities: Science, secularism, and religion in modern India. *Social Text*, *18*(3), 67–86.

Subramaniam, B. (2019). *Holy science: The biopolitics of Hindu nationalism*. Seattle: University of Washington Press.

Tanner, T., & Allouche, J. (2011). Towards a new political economy of climate change and development. *IDS Bulletin*, *42*(3), 1–14.

Thaker, J., & Leiserowitz, A. (2014). Shifting discourses of climate change in India. *Climatic Change*, *123*(2), 107–119.

Yenneti, K., Day, R., & Golubchikov, O. (2016). Spatial justice and the land politics of renewables: Dispossessing vulnerable communities through solar energy mega-projects. *Geoforum; Journal of Physical, Human, and Regional Geosciences*, *76*, 90–99.

Zalasiewicz, J., Waters, C. N., Williams, M., Barnosky, A. D., Cearreta, A., Crutzen, P., … Haff, P. K. (2015). When did the Anthropocene begin? A mid-twentieth century boundary level is stratigraphically optimal. *Quaternary International*, *383*, 196–203.

OPEN ACCESS

# Contradictions of citizenship and environmental politics in the Arabian littoral

Ian R. Simpson

**ABSTRACT**

This article scrutinizes the limitations of environmental citizenship among citizens and non-citizens in the Arab Gulf states, with a focus on the United Arab Emirates (UAE). There are particularly heightened concerns about water scarcity, food security, marine pollution, and dependence on oil and gas industries and how states can address these challenges in the Gulf Cooperation Council (GCC). Yet environmental citizenship in the Indian Ocean's Arabian littoral remains poorly understood both in terms of theoretical and grounded questions. This article considers how labor relations and discourses relating to citizenship, environment and sustainability enable or foreclose environmental reform in GCC countries. It shifts the technological and economic focus predominant in literature on sustainability in the GCC to take a more social perspective and examine distinctions between citizens and non-citizens and the depoliticising of environmental claims and national industrial legacies.

## Introduction

This article scrutinizes the limitations of environmental citizenship among citizens and non-citizens in the Arab Gulf states, with a focus on the United Arab Emirates (UAE). The UAE and Oman are currently the only Arab Gulf members of the Indian Ocean Rim Association, but other Gulf Cooperation Council (GCC) states, notably Saudi Arabia and Qatar, also have significant transnational ties and increasing investments with countries across the Indian Ocean.

Focusing on transnational migration and labor in the Indian Ocean within a framework of environmental citizenship in the Gulf Cooperation Council (GCC) helps to illuminate ecological and environmental aspects of these economies that are otherwise largely overlooked. Employing a political ecology and critical heritage approach, I consider how labor relations and discourses relating to citizenship, environment and sustainability enable or foreclose energy and environmental reform in the GCC countries, with the aim of providing a conceptual and grounded problematisation of environmental citizenship in this regional context. The article questions the ideal notion of environmental citizenship often applied and discusses the challenges to this ideal in a social, political and historical framework — rather than as scientific and technological problems alone — focusing on the significance of migrant communities and transnationalism in GCC countries. Based on this framework, I argue that the distinction and specific relationship between citizens and non-citizens in GCC countries diminishes pressures for environmental reform due to a culture of citizen privilege built on extractivism, the social and political exclusion of the expatriate and migrant majority, and the occluding and depoliticising of environmental claims and hydrocarbon industrial legacies.

In recent decades GCC states and non-state actors have begun implementing policies and efforts aimed at public awareness and action in relation to environmental issues and initiating more open public debate about environmental challenges. However, despite recent efforts to place sustainability in GCC countries in critical perspective (Sillitoe, 2014), there is still relatively little general or contextualized social analysis of conceptions of nature, environment and sustainability in the region. The article therefore attempts to unpack the epistemic and normative dimensions of citizenship and environmental governance in the Arab Gulf, with a focus on the privileges of citizen status, labor relations and constructions of belonging and sustainability in national heritage, with particular attention to contradictions in the relationship between citizens and expatriate or migrant groups in GCC societies.

During the twentieth century, Arab Gulf states relied extensively on immigration and migrant labor to build extractive industrial infrastructures, develop petro-economies, and subsequently to work in education, health, IT, consumer commerce, mass tourism, leisure and other sectors that indirectly resulted. Today, dependence on very large expatriate workforces is deeply embedded, political participation is generally weak, efforts to nationalize the workforce are resisted, and rigid social, political and economic boundaries are often drawn along national lines between communities. The possibility of duties and rights and their social outcomes in GCC countries have to be considered within these unique demographic, social and political realities, and, as I discuss below, environmental citizenship needs to be understood in these contexts not as bounded by the nation state but as global, regional or transnational in relation to belonging, place-making, rights and duties. I show that the complexity of scale involved due to multiple types of migration and transnational relations means that environmental citizenship must be conceived as more spatially distributed and in historical perspective in order to grasp the entangled ecologies and livelihoods that transnationalism and globalized extractivism entail.

The article first discusses the environmental significance of endowed citizens and migrant laborers and then the relation between these two categories, based on which I show how their interrelations work to uphold GCC cultures of consumerism as well as a structure of social-political exclusion. I then turn to discuss environmental discourses in governance and cultural constructions of belonging and sustainability, showing how these are also related to the distinction between the citizen-foreigner relationship in the GCC region and how they preclude environmental citizenship in terms of questioning existing human-environment relations embedded in industrial extractivism and in terms of debating the rights of migrant communities, despite the GCC simultaneously depending on this majority as the labor force that has designed, built and serviced urban projects, industrial infrastructures and tourism in the region, which in turn enable consumption. The article can therefore be read as a problematisation and a grounded argument about national and transnational belonging, migrant labor and environmental citizenship in a specific setting that has arisen from extractivism, transnationalism and globalization.

## Endowing citizens, limiting duties

The question of environmental duties and rights in the Arabian Peninsula can be approached by addressing how it applies differently to each of the two main structural groups, national citizens and expatriates or migrants, and their contrasting social contracts with the rulers and governments of GCC countries, although the duties and rights of these two groups are interrelated in the sense that each group's economic and social status leads to the concerns and limitations attached to each other's duties and rights. I will address national citizens first, arguing that state initiatives aim to persuade citizens to pay for environmentalism based on an assumption of individual duty that is in conflict with privileges characteristic of GCC citizenship.

In the Arab Gulf states, the distribution of capital derived from fossil-fuel extraction is central to the differences created between tribal, elite and other citizen families on the one hand and non-citizen social groups on the other (Gardner, 2010; Kapiszewski, 2001). National citizens are endowed through state welfare spending, benefits and public sector salaries by a ruling monarchy that redistributes government revenues (corporate tax revenues on oil and gas) from the surplus economy, while other (non-citizen) groups acquire capital predominantly through the means of labor. In general there are no taxes on wage income for national citizens or for non-citizens. Citizens of GCC countries have been afforded direct handouts or subsidies and given the status of landholders with the authority to employ people, upon which the system of migrant labor hinges. It is significant that national citizens are the only people with the official authority to sponsor and employ guest workers. Paul Dresch (2005; 2006, pp. 203–204) has discussed concerns about guarding citizenship and its privileges in the face of very high proportions of foreign residents in Arab Gulf states, reflected by restrictions on marrying foreigners and descent defined within rigid boundaries. The elite groups' association with state government and their kin relations are also maintained through a specific way of controlling capital and labor, namely by relying on a political economy and decidedly liberal organization of labor that excludes the majority from most of the state's provisions. Citizens are also privileged by access to the vast majority of employment in the state public sector, although these jobs are distributed unevenly along family, sectarian and tribal lines (Gardner, 2010, pp. 144–146), while non-nationals nearly all work in private sectors. Another aspect of the relationship to the state that should be emphasised is that political representation and participation is generally weak.

The above dynamic and its specific possibilities and limitations that influence environmental ethics, policy and action in the Arab Gulf states will be explored in more detail, such as the effect of endowed citizens on their perception of environmental duties, and the relationship between political participation and social outcomes concerning environmental issues. In most Arab Gulf states where national citizens make up only a minority among a majority of diverse expatriate groups, it is worth contemplating to what extent national citizens form a body politic or socially coherent community and whether this is a factor in the ability to debate, cultivate and organize around a new set of environmental ethics within the privileges of national citizenship.

The political economy of GCC countries raises issues about the effectiveness of both citizens' requests for government policy and action regarding issues such as pollution and global warming, and the government's ability to implement change requiring private individuals to act. For example, when it comes to large-scale energy transition and installing solar power at the individual household level, Luciani (2014, p. 21) notes that

> encouraging small investors such as families or commercial enterprises to exploit their roof space or cover their parking lots with PV panels requires a more complex and difficult relationship between the single buyer and a very large number of potential providers; this kind of relationship is unusual in the region, where individual citizens and families have only sporadic direct contact with government authorities and the administrative capacity of the state to penetrate society is limited.

In addition, the provision of highly subsidized fuel, electricity and water for homes and personal cars makes it especially difficult to incentivise or change 'consumer decisions and behaviours towards less wasteful use of resources', making large, fuel-inefficient vehicles 'a desirable and economically viable alternative' for instance (EWS-WWF, 2015, p. 69).

Thus endowed or privileged citizens and a lack of political participation (notwithstanding other, local cultural forms of participation) limits rights and obligations of citizens with respect to environmental issues. Citizens are treated primarily as consumers of environmental goods and services, yet citizens enjoy subsidies leaving little incentive to change consumption or to debate existing and alternative environmental ethics. The green economy, sustainability and energy transition in the Gulf is mainly approached through a contractual view of environmental responsibility in which the government and citizens are the focus, and the challenge of greener economies and greener living is addressed as a problem of technology and the skilled labor this requires. This notion of local, contractual environmental responsibility is deployed despite the relationship between the citizen and the 'state', or ruling family governments, in GCC countries where citizens have not paid the taxes normally thought to impose obligations on the state towards them (although recently governments have attempted to introduce Value Added Tax and other taxes). In light of this particular social contract in GCC states, what is the potential of the relationships between governmental authorities and social groups in engendering debate, action and reform on issues related to energy, environment and nature?

There is a growing number of empirical studies in Arab Gulf contexts on evidence for 'environmental citizenship', even if they do not employ this term explicitly. The studies aim to establish what citizens value, their attitudes and so on, with the aim of understanding how policy and efforts to instill environmental responsibility among citizens can be best formulated. For example, a Willingness to Pay study in the UAE conducted in 2014–15 uses survey data from a sample consisting of 1,282 citizens and indicates that respondents perceived environmental problems not as individual but as a government social responsibility (Yaghi & Alibeli, 2017). The authors argue that citizens are overly dependent on government for environmental protection. While the Willingness to Pay methodology is used to explore how people perceive environmental problems, it presupposes environmental citizenship in terms of ideal state and non-state actors. The authors contend that citizens in the UAE need to be made more aware that they have a major role to play as engaged and responsible citizens, that such realization will lead to a willingness to pay to protect the environment, and ought to shoulder this burden by paying environmental taxes. As Bell (2014, p. 357) asserts, this is environmental citizenship made 'complicit in the neoliberal agenda of privatizing and individualizing responsibility for environmental problems that can only be solved by collective political, and ultimately state, action'.

This thinking is in line with the UAE government's New Public Management approach which 'targets people's policy engagement by emphasizing their moral obligation to act upon those environmental values in which they believe', and assumes that such change equates to values of equality and fairness in society (Yaghi & Alibeli, 2017, p. 53). Here we see an assumption of duty of the individual, which, despite claims that government and citizens be involved in open dialogue, presupposes the citizen is a normative actor that has environmental responsibilities, rather than problematizing the situation by starting from the contradictions or paradox of the current composite of a political economy of rapid growth, extensive transnational workforces, consumption, GHG emissions and so on. It is clear these studies and the state discourse they engage with reflect a tendency to instrumentalise citizenship in order to foster greener behavior and sustainable development, in which an instrumental sensibility 'takes the focus off citizenship as an actively (and not merely theoretically) contested concept/status/practice' (Latta, 2007, p. 385).

An interesting aspect of Arab Gulf states promoting environmental responsibility among citizens is the assumed link this approach has with efforts to move away from extensive welfare and subsidies for citizens, introduce taxes and diversify GCC economies. Again, rather than considering less consumption and less growth as a starting point to reduce and mitigate environmental problems, the idea is to create citizens that pay for being pro-environmental. Others criticise such demands as too narrow and unlikely to result in greener behavior without 'broadening environmental citizenship' by fostering an overall architecture or culture of greener, environmental thinking and acting (Horton, 2006). While painting a broad picture as I have here can eclipse variation and different approaches, the GCC is generally a social and political landscape where citizenship, with its privileges of state handouts and zero taxes, has been built on revenues made from unmatched fossil-fuel sources and where environmental citizenship is not seen as an opportunity for nationals to question and rethink the nexus of extractivism, petro-economies and environmental crises in which they live but to reframe subsidies for citizens as services that they should feel obligated to contribute to financing, including environmental regulations and protection.

## Migrants, rights and embodied materialities

Turning to expatriates and migrants, the other main structural group, this section shows how the possibilities of environmental movements are also limited by the social and political exclusion of expatriates and migrants. It is remarkable that although most GCC countries have comparatively small populations — for example 9.63 million in the UAE, 4.83 million in Oman, but larger in Saudi Arabia with 33.70 million (numbers in 2018 World Bank, 2019, June 29) — they are host to high numbers of non-nationals who disproportionately comprise the total labor force, mainly in construction and domestic work but increasingly in other sectors and at all 'skill levels', and non-nationals make up the majority of the population in most Gulf countries. For instance, in 2017 expatriates comprised 45% of the population in Oman and 75% of the population in Bawshar province where the capital Muscat is located (National Center for Statistics and Information, 2018; 2019, June 29). Expatriates comprised 88.5% of the population in the UAE in 2010 (UAE Government, 2019, June 29), and 92% of the Emirate of Dubai in 2018 (Dubai Statistics Center, 2018). Moreover, migrant workers constitute even greater proportions of actual labor forces in GCC nations (Al-Ubaydli, 2015), and figures for some districts are likely to be underestimates due to undocumented, unofficial and forced-labour migrants and refugees. Saudi Arabia and the UAE are, respectively, the fourth and fifth largest migrant countries globally and the GCC nations together host more than 10% of all migrants worldwide (ILO, 2019a, June 29; 2019b, June 29).

Overall migrant figures alone say nothing of the precarious nature of much migrant labor, the wide diversity of communities, transnational connections, and hierarchies within them, or changing trends in migration and mobility. For instance, in the 1990s, the largest migration to the Arab Gulf ensued from India, and especially from the state of Kerala and from certain districts within Kerala (Willoughby 2006). Out-migration from India still predominates but recently the number of migrants from wealthier states like Kerala has declined and migration from states in northern India is growing (Chanda & Gupta, 2018). Another trend has been a shift from only a small proportion of female migrants in the 1990s to double the number by 2013 (Chanda & Gupta, 2018).

These hugely-significant expatriate labor forces have been encouraged through open entrance requirements and private employment enabled through the infamous *kafala* system of sponsorship, in which a national citizen, or organization, has the right to employ people and is also meant to be liable for their conduct and safety (duty of care). Kafala governs the legal entry and status of expatriates and migrants and allows employees to work only for their sponsor. Scholars have argued that kafala, originally a work-permit system in the rentier-state paradigm, has evolved into a foundational, controlling platform of exclusion in society that is a consequence of state-citizens endowed with privileges and authority over patronage of the majority of people in the GCC region, their access to labor, and so on (Dito, 2015; Longva, 1997). Since the 1960s, migrants became separated from the resident population

> not only by the national and ethnic re-composition and diversification of the residential and working population, but also through the introduction of more developed forms of geographical, workplace and legal segregation and rotation designed to minimize the possibility of political, economic, cultural or social links being forged between migrants and the indigenous population. (Chalcraft, 2010, p. 22)

Crucially, maintaining state capital only among patrons serves as an impulse to preserve kafala because it provides the labor mechanism that partly enables production, the revenues derived therefrom and Gulf cultures of consumption.

The kafala system can be very restrictive for expatriates' rights, labor contracts and freedom of movement, as the International Labor Organization sums up: 'While in some cases, employers may welcome the responsibilities of the kafala system and treat the worker well, the inherent imbalance in the rights and responsibilities of each party can create a situation which is exploitative of the worker' (ILO, 2019b, June 29). If the tendency for abuse and exploitation of migrants in the GCC is widely attributed to the system of sponsorship on which migrant labor operates (Fargues & Shah, 2018), the possibility of attuning and attending to its associated ecological and environmental challenges initially seems meagre when this system is widespread and underpins the majority of the Gulf's workforce. The kafeel status of citizens is a double-edged privilege, both obligating responsibilities to guests and favoring their exploitation. Kafala shifts environmental duties from the state to the citizen and thereby responsibility can tend to be diffused. Political exclusion due to lack of civic and political participation of non-nationals in the Gulf (or their place of origin) does not allow representation of environmental matters among the GCC majority populous (expatriates), yet if environmental concerns like exposure to harmful conditions are to be addressed, let alone other changes, then recognizing environmental costs, political participation and efforts to nationalize workforces will be needed.

So far I have touched on citizens and the non-citizen majority separately, but this distinction is important in many walks of life in the Gulf not only as a distinction but, I argue here, as a dialectal relationship both in principle and in grounded ways: in the structure of GCC societies, nationals and expatriates/migrants enable each other's relative position, status and environmental burdens in a fundamental way. Dresch (2006, pp. 200–201) has argued for conceiving the structure of these societies as specialized polities in which the distinction between national or citizen and foreigner (expatriate or migrant) is integral and the non-citizen is conceived not as an issue of controlling migration and borders, but as constitutive of society in a way that is particular to GCC states. Extending this idea, the subjects created by GCC nations (whether citizens, camel races, tall buildings or shopping malls) are only enabled by foreign, non-nationals living in the region, and yet it is national subjects that have often drawn the attention and predominate in studies of culture and nationalism in GCC countries because they appear as iconic of national efforts alone. I will return to the citizen and the foreigner (the expatriate or migrant) and argue that rather than being a dichotomy, they are part and parcel of environmental citizenship and its constraints in the deterritorialised, transnational spaces of the Arab Gulf.

I have discussed citizenship and the limited rights and participation of migrant communities in GCC states in order to sketch out how these relations might be important for situating any notion of environmental citizenship. Due to the above social stratification and divisions, conditions and challenges are experienced differently by different groups in the context of the harsh and sensitive ecological environments of the Gulf. For instance, the embodiment of coping in the region's arid to very arid climate, extensive sun exposure and very high seasonal temperatures and humidity levels can vary strikingly between citizens and migrant communities, and needs to be given attention. Mean air temperatures in Abu Dhabi for the period 2003–2018, on aggregate, ranged from 20.1°C in January to 35.5° C in August, reaching a mean maximum of 40.2°C and an absolute maximum of 48°C (National Centre of Meteorology, 2019). In Muscat the minimum/maximum temperature in January was 16.7/25.1°C, climbing to 30.3/40°C in June, based on mean values observed over a 30-year period (World Meteorological Organisation, 2019, June 29). Notably in 2018, Oman recorded its highest ever *minimum* temperature (42.6°C) to date, in Quriyat (NOAA National Centers for Environmental Information, 2018). On land, evaporation is high, annual rainfall typically of less than 100 millimeters and vegetation cover sparse (Cadène & Dumortier, 2013). Prevailing winds (the *Shamal*) on the south coast of the

Gulf increase humidity levels as they bring moisture ashore. It is the combination of very humid and hot air—best indicated by *wet bulb* temperatures—that places human and non-human bodies at their physiological limits, and migrants working in construction, domestic jobs, factories and other sectors may find themselves in these conditions. Many live and work without air-conditioning, while in contrast citizens are mainly afforded the comfort of artificially-cooled office workplaces, homes and vehicles. In some versions, environmental citizenship is conceived as extending rights to future generations of human and non-humans, ensuring them healthy ecosystems and limited climate change. This conception helps to reveal the additional aspects of expatriate experiences, who toil in industries that require exposure to extreme environmental conditions, and to draw attention to those who are politically and structurally excluded and isolated.

Does economic growth need to be so rapid that people in construction work shifts in extreme conditions, rather than regulating the sector so that they do not endure extreme humid-heat conditions during the long summer? It is obviously a duty within the labor contract to ensure people are not exposed to extreme, toxic or unsafe conditions, yet this is neglected precisely because the environmental body politic remains nascent in the GCC. I argue that it is only through accounts of migrants' experiences, such as ethnographic descriptions and stories, that the controlled and surveilled body politic can be glimpsed through a different lens that reveals experience and vulnerability as the body endures in some parts of the CGG migrant labor system. For instance, in the case of the Malayalam novel Goat Days, Benyāmin (2016) vividly portrays the daily sweating and stench that the character Najeeb, a migrant from Kerala, is forced to experience while indentured in farm labor in Saudi Arabia. Mahdavi (2012) retells narratives of domestic migrant workers in Dubai who moved from the formal to the informal economy and the potential abuse this can lead to such as forced labor and trafficking. Bristol-Rhys (2012) describes how migrants experience social boundaries in Abu Dhabi, where laborers live in camps isolated from the city and are ferried between work sites in special buses, while city workers live in rundown parts of the city, often sharing crowded apartments and in some cases taking turns to sleep due to space limitation. Laborers, migrant workers, expats, and Emiratis are also separated by the way other urban spaces in Abu Dhabi are designated or understood as social spaces specific to family, the wealthy or otherwise off-limits for some due to entrance or transport requirements, and migrants are assuredly policed along these social boundaries.

Any account of migrant communities does not simply define them by reference to their subordination but can be read to illuminate their lived experience of the conditions they are put under and to attend to greater possibilities that can emerge from this toward environmental reform. Revealing the political ecologies and hidden environmental issues of large or majority populations of migrants in this way is a necessary step not only toward adopting ethics and regulations that would prevent labor in extreme, toxic or unsafe conditions but also toward a fuller, *inclusive* accounting and knowledge of the overall ecological and health impacts of GCC petro-economies.

## Foreclosing rights and duties

I discuss further here the interrelationship between nationals and non-nationals in the GCC to consider the utility of environmental citizenship in its liberal form by exploring labor relations more in depth. Dresch (2006, pp. 214–215) provides an example of these relations which involves fishermen in the northern UAE and how its particular social structure leads to duties and also objections to migrants and concerns about environmental problems that can be contradictory:

> Some years ago now, having absorbed the language of ecology and conservation, owners of the grounds were in danger of being fished out. The damage was attributed to "Asian fishermen." It is only going through old newspaper files that one realizes, first, how long this has been going on and how often a law has been passed that has then not been applied, and second that the "Asian fishermen" in question are actually the employees of citizens. It could not be otherwise. They are not sailing over from the Malabar Coast or the Comoros Islands but are resident on the Gulf coast and therefore have local *kufalā'* or "sponsors." Finally a law was passed (Federal Law No. 23 of 1999) insisting that local owners be present on these boats at sea to supervise what was done; the owners, having denounced the generalized Asian presence at length, then protested volubly that they had other things to do and could not spare time to bob around on the ocean supervising individual crews.

This localized example is symptomatic of a broader contradiction integral to developing a contextualized practical understanding of environmental citizenship in the UAE and other GCC countries: considering how the duties of national citizens, due to the sponsorship system, extend to a responsibility for the activities and actions of people employed under their privilege as citizens. Since GCC citizens carry the state-like authority to grant entry and employment to potential employees, the kafala system raises the question of whether it therefore also imposes on *kafeel*-citizens responsibilities concerning the environmental and ecological aspects of labor in these countries.

Assuming economic, civil and political rights include environmental rights, this labor relationship with non-citizens in GCC countries can thus serve as an indicator of whether the environmental experience and rights of expatriates are addressed. For instance, Pandya and Sparding (2016) have recently discussed labor issues in the context of the 'green economy' in the GCC, including how the requirements of a green economy shape labor migration and vice versa, especially from the perspective of institutions, employers and the International Labor Organization. While the authors focus on the skilled labor force needed and obstacles to creating it, the obstacles they identify are relevant to imagining any practical application of environmental citizenship in its liberal version. Pandya and Sparding (2016, pp. 197–198) argue that

> whereas social inclusion has now become a standard feature of the international discourse on environmental protection, green economy and green growth, the specifically labour dimensions of social inclusion will face significant obstacles in many of the GCC countries. Trade unions and other workers' organisations are weak or non-existent in Qatar, Saudi Arabia and the UAE, beleaguered even in Bahrain where they previously enjoyed some influence with policymakers, and nascent in Oman. In the absence of strong labour voices, the capacity and willingness of policymakers to attend to workforce considerations from the perspective of workers' interests will thus be slight.

In other words, the labor system leads to the relative social exclusion of migrant workers in GCC nations and also forms the obstacle to recognizing workers' environmental rights generally, not only those of migrant workers, thus having an impact more broadly. In a system that exploits transnational inequalities in labor for the needs of production, it is difficult but necessary to reveal and act on the hidden environmental costs that are avoided by politically excluding and structurally isolating large numbers of migrant workers. For instance, how can we apprehend the ecologically unjust differences between citizens and migrant workers living without decent conditions? I propose that it is by unpacking the structure of the political and labor relations enabled through transnationalism in the GCC economies that we can grasp how this creates the space for citizens and governments to feel they are making a real shift to more sustainable behavior (citizens doing their bit to consume more sustainably, and governments making services more environmentally efficient and lower-carbon), but to still rely on a politically underrepresented or unrepresented workforce to supply those changes. It is through the lens of labor that we can see this discordance, at least in the affairs of GCC nations, where in practice environmental citizenship appears structurally limited when presented with questions such as whether national citizens would feel responsible for creating a green economy, or green jobs, when these countries rely principally on a permanent flow of temporary migrants to meet workforce demands.

The material and spatial dimension of the citizen-foreigner distinction enabled through GCC-configured labor and transnationalism can also be examined to reflect critically on environmental citizenship. The materiality and spatiality of different social groups — such as the large villas owned by national citizens and staffed by foreigners, houses rented by Europeans and other migrants, and labor camps predominantly for Asian migrants in Dubai — need to be accounted for in studies of urban environmental relations and the possibilities and limitations of environmental citizenship. For example, Dresch (2006, p. 210) reported that for Emiratis in Abu Dhabi it is considered respectable to live in a private villa with an outer enclosure, which is associated with security and independence, and deviation from this can be seen as weakening cultural and social ties to other nationals. Keeping in mind the close guarding and rigidity of GCC citizenship, if the walled villa form is partly constitutive of being a citizen, it means taking seriously the political ecology of home that has emerged in the citizen-foreigner distinction as part of the structural possibilities and limitations of environmental belonging or citizenship. It becomes important to reveal how the water needed for villa gardens and the numerous foreign domestic staff needed to tend them are part of and integral to the ecological aspects of citizenship in the UAE, if not the other Arab Gulf states as well. The Emirati villas contrast with the mass camps for migrant laborers, which by comparison represent a low environmental burden in consumption terms but a high health risk with their overcrowding and poor conditions, and yet migrants intentionally hide these conditions from their home family and instead attempt to stage a consumerist lifestyle (Gardner, 2012). The villa gardens in some way parallel an official, national strategy of greening the UAE (Ouis, 2002), and when it comes to tourism this greening and the meanings of beauty, vitality and modern progress that greening aims to symbolize awkwardly conceals the ecological context of aridity, water scarcity and impacts of tourism development, especially when combined with the unfettered rights to consumption given to tourists (Stephenson & Vyas, 2017). Such questions about disjunctures between environmental attitudes and realities (Robbins & Sharp, 2003), which start with understanding patterns of migration, ethnicity, labor relations and social and economic status, are as crucial in Indian Ocean settings as they in any context.

Another strand of the environmental dimension of transnational labor in the GCC is in the role of remittances, which can be significant in environmental protection in places of origin. Campbell (2014, p. 213) points out that sources of labor for Nepalese migrant workers in Qatar generate remittances that support distant conservation initiatives and protection of biodiversity in other places, such as in Nepal, and with good reason highlights the need to study connections between sustainable biodiversity conservation and migrant labor. Similarly, we can ask more generally how the globalizing forces that have expanded transnational mobility and migrant labor have varying, indirect ecological effects in distant places.

The globalizing dimension of migration is clearly in additional tension with liberal versions of environmental citizenship in GCC countries not only because of the practical difficulty of citizens acknowledging a duty to distant strangers in other territories that have caused far smaller ecological impacts (Dobson, 2003), but also acknowledging a duty to their expatriate populations within Arab Gulf countries. This aligns with and extends Hailwood's (2005) point about needing to account for the stranger in environmental citizenship, including transnational migrant workers and their distant locales of return. It seems increasingly important theoretically and practically as migration, displacement and mobility are driven, in part, by the effects of global warming and environmental problems. The citizen and the stranger, rather than a dichotomy, are part and parcel of the difficulties in discerning the duties and rights of environmental citizenship at multiple scales within Arab Gulf states.

## Environmental discourses and initiatives

The ecological and environmental relationships entrenched in GCC oil and gas societies have received comparatively little attention in social science and political science studies. In contrast, this section takes a broad look at policy and discourse of environmental authorities and private industry to reflect on how GCC governments and non-state actors have more recently begun formulating and implementing policies and efforts aimed at raising awareness and taking action on issues in the context of energy transition, issues of food and water security, and governments in the Gulf engaging more openly in public debate about environmental concerns.

Cities in the GCC face potentially high degrees of social vulnerability and impact due to climate change, pollution and related shortages of energy, food and water. There are particularly heightened concerns about sustainability due to high consumption levels, dependence on imports and arid environmental settings in GCC countries. Water scarcity is already a pressing issue (Ward & Ruckstuhl, 2017), exacerbated by rising demand as agriculture and urban populations expand. Industry, urbanization, tourism development and coastal reclamation have adversely affected the marine environs, for instance by affecting fauna such as mangrove plants that support sea animals. Further problems are industrial pollution (oiling of wildlife, insufficiently treated wastes, discharges from desalination plants, and pollution from shipping such as oil spillage and dumping of ballast) causing indirect impact on ecosystems and shallow-water marine species vulnerable to temperature variation due to anthropogenic global warming (Abuzinada, Barth, Krupp, Böer, & Abdessalaam, 2008; Khan, Munawar, & Price, 2002; Krupp & Abuzinada, 2008; Sheppard et al., 2010; Shriadah, 1999; Stephenson & Vyas, 2017).

While the GCC has primarily been a major oil and gas exporting region, its demographic and industrial growth that has indirectly resulted has also driven up energy consumption in Arab Gulf countries to among the highest per capita worldwide. This development has in turn created increasing reliance on petroleum and natural gas in order to produce energy for water desalination and air-conditioning, among other uses. The rapid and continuing expansion of industry, population and urban areas, mass tourism and rising consumption levels form part of a wider set of external and domestic factors to which Arab Gulf states are responding (Fox, Mourtada-Sabbah, & al-Mutawa, 2006). If the petro-monarchies of the Arab Gulf are emblematic of a globalized ideology of extractivism and growth, is it also those governing elites that should lead the way in cultivating new values and implementing fundamentally different ideals under an energy transition, and to what extent are they playing this role?

Although GCC countries rely deeply on oil and gas extraction and are among the highest per-capita energy consumers globally, all six GCC countries have ratified the Paris Agreement, Oman most recently in 2019, and are making efforts to introduce policies and action in line with the United Nations Framework Convention on Climate Change (UNFCCC). Taking the UAE as a notable example, in recent years it has perhaps been the most prominent in taking steps to put in place authorities and policies concerned with energy transition, climate change, and sustainability. In 2015, it launched the UAE Green Agenda 2015–2030, and in 2016, established a Ministry of Climate Change and Environment and the UAE Council on Climate Change and Environment. In 2017, the UAE formulated its Energy Strategy 2050, which aims to increase clean energy sources to 50% and reduce the carbon footprint of energy production by 70% by 2050, and drew up the National Climate Change Plan for 2017–2050, which aims to manage greenhouse gas emissions, increase climate resilience and promote economic diversification, all while sustaining growth (Ministry of Climate Change and Environment, 2017). Well known among the numerous organizations that have partnered with the UAE government on environmental and sustainability challenges are The Sustainable City in Dubai (https://www.thesustainablecity.ae/), Masdar City (Günel, 2019) and the related Masdar Institute (https://www.ku.ac.ae/institute/masdar-institute/), which is now part of Khalifa University, and Emirates Wildlife Society in association with World Wildlife Fund (EWS-WWF, 2018). These government-related organizations and the government itself are endorsed by an absolute monarchy and its contemporary tribal leadership. The International Renewable Energy Agency is also currently headquartered in the country, and the UAE has altogether positioned itself as a regional leader in efforts to innovate sustainability and decarbonization of its cities and energy generation.

The Masdar City initiative is a good example of the extensive yet concealed or overlooked reliance on migrant labor in developing a sustainable eco-city that would shift dependence from petroleum revenues in the UAE and 'liberate humanity from its guilt-ridden consciousness of the twentieth century' (Günel, 2019, p. 63). Günel argues that inserting existing divisions of labor and social space into Masdar City and relying on temporary migrants to construct and maintain the site have mattered less than valuing and developing a vision for a future of sustainable living and renewable energy.

These developments are paralleled by studies, working papers and grey literature that typify an empirical assumption that organizations and states in the Arab Gulf do indeed function as ideal environmental actors. For instance, Alam and Luomi (2018) suggest that GCC governments would best facilitate non-state and subnational actors on climate action through means of state consultation, considering the otherwise limited political participation and avenues to exercise political action. They advocate for governments raising awareness about 'the benefits of measuring and managing GHG emissions, and of energy efficiency and renewable energy as cost-effective mitigation measures' (Alam & Luomi, 2018, p. 34), with the aim of encouraging non-state actors to join transnational climate governance initiatives (initiatives that are governmental efforts to address climate change through transnational means involving at least one non-nation-state actor by explicitly seeking to govern (steer or conduct) a constituency towards specific public ends, as defined in Bulkeley et al. (2014, p. 18)).

There is a strong focus on sustainability in a range of issues, but in most cases this is a normative, ideal notion of sustainability. The environmental duties and rights framework of citizenship is used to promote a discourse of 'technological' sustainability but not to scrutinize the extractivism basis of GCC petro-cultures. Also lacking is a critical view of sustainability that addresses issues of justice and equity in relation to citizens, non-citizens and transnational connections, for instance, of the social and economic inequities encouraged by the differential environmental conditions experienced and the structure of migrant labor relations in the GCC. In this sense the discourse on sustainability and environment in the GCC often lacks a realistic inner landscape of society in these countries. The discussions of sustainability, greening, energy transition and other environmental challenges in the GCC (Abdel-Raouf & Luomi, 2016; Abdelraouf & Azar, 2017; Akhonbay, 2018; Bryde, Mouzughi, & Al Rasheed, 2015; Luciani & Ferroukhi, 2014; Luomi, 2015; Sillitoe, 2014) in the main lack consideration of the political dimension these issues, particularly an articulation with social inequities in the Gulf such as the politically marginalized majority population and social and economic hierarchies within the labor force. Crucially, GCC regimes of governmentality preclude environmental citizenship in terms of rights for the large migrant populations, despite at the same time depending on this majority to enable production and consumption. The following section draws a link between this area of governance and national constructions of belonging and sustainability.

## Promoting sustainability

Cultural heritage is a field of construction that often shapes national identity and can be used by social and political actors to foster environmental responsibility among organizations and the public. This might be to cultivate alternative approaches and perspectives to consumption, energy and sustainability, but heritage can also be used to legitimate and uphold dominant or normative views of human-environment relations. Here I consider whether heritage practices address the ethos of growth and consumption, or exploit and (re)shape the past and history in the present in attempts to inform sustainability initiatives in productive or romanticizing ways. I also want to show how these efforts connect with, or marginalize, significant numbers of people when they are constrained by the homogenizing tendency of national heritage.

Knowing that nationalism is engineered partly through cultural practices, it would seem reasonable to suggest that just as states have invented traditions, constructed heritage, and fashioned cultural forms in order to define national identity, they can, in principle, also exploit these same tools to forge a more environmentally and ecologically slanted discourse about the nation, citizenship and belonging. Looking at ways organizations mobilize and deploy tradition in their attempts to foster and shape environmental citizenship, the example here regards construction and promotion of the former pearling industry in the Arab Gulf and awareness campaigns and community programs relating to food and resources in the context of present concerns about sustainability. It also concerns who these practices are for, the exclusion of historic and contemporary migrant communities, and the question of political participation.

Environmental organizations in the GCC are involved in efforts to influence consumer choice by issuing consumer guides about sustainable species, some of which recall former human-environment relationships in pearling and other livelihoods. For example, Emirates Wildlife Society works extensively at raising awareness and promoting sustainable lifestyles through direct public campaigns like the Choose Wisely Campaign concerning fish stocks and sustainable fisheries, the Heroes of the UAE program to save energy and water in the home and workplace, and other programs and guidance to inform consumers about energy-efficient and water-efficient products and greener food choices (see EWS-WWF, 2015). These efforts come in the context of degradation or loss of habitats along the coastline of the Gulf from Kuwait to Oman threatening marine species due to the human-environment relations of fossil-fuel extraction, commercial fishing and unintentional catch, urbanization and mass tourism.

Direct outreach to the public as a consumer body attempts to produce types of responsibility among consumers, and this work takes place within a wider set of environmental efforts, such as the Environment Agency Abu Dhabi's monitoring of the fishing industry and raising the awareness needed for conservation of marine species. The EWS-WWF Choose Wisely Campaign harbors a notion of valued 'natural' habitats and ecological knowledge that are seen as embedded in the heritage of fishing and pearling. Consumers are informed about endangered species of fish in the Gulf and responsible buying choices. EWS-WWF has drawn on the UAE's connection to the sea through reference to its traditional pearling and fishing industries and professes that Emiratis are entrusted with this 'heritage'. Pearling and marine heritage are leveraged to draw attention to environmental concerns, to shape an understanding of human effects on marine habitats, and to engender sustainability in consumer culture (Simpson, 2014). This dimension of conservation and cultural heritage relates to current conversations between anthropology, cultural geography and environmental sciences concerned with grasping the nature of post-industrial ecological change (DeSilvey, 2017; Harrison, 2015; Harvey & Perry, 2015; Tsing, 2015).

Despite organizations actively targeting consumers as ideal citizens, it should be noted that the ecological and environmental setting in the GCC means there is litte option for citizens to reconnect with local food producers or avoid global supply chains, long-distance transportation or continuing to import nearly all food. Former agricultural practices consisted of a combination of pastoral nomadism and irrigated cultivation, while fishing provided fresh fish on the coast and dried fish for exchange. As groups shifted to more sedentary modes of livelihood, GCC states today have implemented industrial farming and invested in overseas agricultural land. The conception of green consumer, whether one that consumes less overall or exercises greener consumer choices (Seyfang, 2005), also comes up against rapid rates of economic growth in the GCC, including ongoing development of infrastructure to meet the demand, or vision, of expanded consumption.

The former pearling industry is indeed mobilized as heritage for broader purposes throughout the Arabian littoral, not only in consumer campaigns, and is certainly one of the most conspicuous aspects of the past that has been selected and promoted to fashion heritage in the present. Pearling is framed in the present as foundational to the economic basis of these coastal societies and it is interesting to consider what effect this heritage work is meant to have in terms of environmental consciousness, what view of nature and society it intends to convey within the context of environmental challenges and problems in the present. The human–environment interactions implied in pearling and seafaring heritage practices conjure an intimate encounter between humans and the environs in the Gulf and a specific cultural, ecological adaptation. Considering this heritage work in the context of environmental concerns described earlier, it would appear to underline the differences between former livelihoods involving close interaction with sensitive ecosystems — even if viewed romantically — and today's rapidly expanding urban centers and industrial areas, overconsumption and high-carbon energy generation, without however aspiring to address directly or fundamentally the ethos of growth and consumption. The differential leveraging of pearling as the iconic national heritage in contrast to petroleum relates respectively to green discourses and environmentalism on the one hand and toxicity, depletion and global warming on the other. One interpretation of this heritage work is that it encourages people to question terms such as natural 'reserves' and natural 'resources', and to debate what their relationship is with nature in the Gulf today (Simpson, 2014). At the same time, as a dominant heritage discourse and practice, its excluding effect is to banish the materiality of twentieth-century development, such as pollution and toxicity, from what is officially considered as heritage.

So various actors mobilize pearling heritage to engender emergent cultural and social imperatives about the environment in the Gulf. Although it is not the case that environmental policy directly drives and shapes these constructions of cultural heritage, there is scope to do so in ways that could contribute to engendering an ethics of care and critically questioning human-environment relations in the Gulf and environmental responsibility, but it will be limited in who this is for because current heritage practices tend to speak only to national citizens, who are a minority in most of the GCC. Besides its distinctive ecological dimension, this is the other aspect of cultural heritage in the Arab Gulf littoral that I wish to scrutinize. When national heritage is constructed, who does it naturalize as national and as having environmental belonging, and into what place does it cast others, such as the Indian Ocean majority in the GCC?

When it comes to the Arab Gulf's Indian Ocean connections, there is a discourse of modernization (see for instance Fox et al., 2006) in which the emergence of transnationalism in the Arabian Peninsula is marked by oil exploitation and globalization in the second half of the twentieth century, before which the Arab-Persian Gulf is framed as a formerly unchanging, traditional society. Yet for centuries prior to this, urban free-ports on the Arabian coast had traded agricultural goods and livestock to India and attracted people from South Asia and East Africa. There is a history of intermingling in the Gulf's outward facing communities (Al-Rasheed, 2005), which formed in a wider inter-Asian context through historic connections such as between port towns in the Arab Gulf and the Indian Ocean. This resulted in the formation of significant social groups, such as Jews (Poirine, 2003) and Africans indentured as pearl divers in Qatar (Hopper, 2015) in Arabian port towns living alongside merchant entrepreneurs and Arab tribal capitalists who established and operated free-ports in the Arab-Persian Gulf (Fattah, 1997). Personal accounts of migration and patronage, which are still handed down from one generation to the other in the Arab Gulf, reflect the process by which these groups and individuals carved out social spaces. In addition, a strong scholarly focus on the impact of the West, especially British control of the Arab Gulf, has overshadowed these historical transnational connections.

Tracing the contours of this historical and contemporary context of Arabian Gulf communities helps in building the necessary framework to provide a more critical perspective of citizenship in a broader sense in Arab Gulf states and the varying rights and duties regarding the environment in the Indian Ocean today. A problematisation of the social significance of environmental challenges or environmental citizenship in the United Arab Emirates, which does not embrace the historical formation of communities through their Persian, Indian, Jewish or African connections cannot fully explain the limitations of any ideal notions of environmental citizenship in the present.

In the field of cultural heritage in the GCC, there is generally little portrayal of how the above regional and long distance mobility and trade played a role in the social mixing and formation of pre-modern Gulf communities. For instance, notable for their absence are stories of how merchants from Iran and India and laborers from East Africa came to live permanently in the Arabian Peninsula and form part of the social and ethnic inheritance among national citizenry in Arab Gulf states today (Simpson, 2014, p. 40). While there are exceptions, such as an historical account of Indian merchants in the business class in Dubai (Vivekanand, 2008), in general heritage practices marginalize the trans–regional social intermingling in the Gulf, possibly because of concerns, as Dresch (2006) has discussed, about guarding citizenship and its privileges in the face of innumerable foreign residents in Arab Gulf nations. Khalaf (2000, p. 244) argues that inventing traditions such as camel racing also serve to safeguard 'cultural heritage and national identity, as well as the traditional dynastic political structure of the UAE as a modernizing nation-state'. This might also be why pearling heritage is mobilized in a way that makes it the citizen's heritage and not the migrants', a secure nationalistic project rather than a transnational or multicultural nationalizing project that reflects the past and present Indian Ocean entanglements of Arab Gulf societies. This could create a barrier in using heritage practices to foster the broader environmental movements necessary to foster inclusive environmental reforms in the GCC, since excluding or marginalizing the Gulf's Indian Ocean entanglements precludes the Gulf's majority communities. The construction of national heritage and its corresponding citizenry plays a role in limiting environmental citizenship in practice, which is bound up with the structuring of society around the internal citizen-stranger relationship discussed earlier.

## Contradictions of citizenship

Organizations referred to above in the GCC have too simplistically adopted a conception of liberal citizenship and civil society without considering the context of Arab Gulf states, their distinct state-society relations and the presence of large numbers of migrants in the mix, and in some places tourists too. This has contributed to making the application of environmental citizenship appear awkward in the context of the significant citizen-foreigner relationship in GCC countries and their specific historical developments, which include the remnants of rentier state policies that were encouraged by Western and international powers.

If expatriates constitute a permanent feature of temporary-labour in GCC nations, the hierarchy of citizen and foreigner and its role in shaping and structuring society needs to be examined to understand the possibilities of environmental reform. The above discussion has identified multiple sites in which the possibility of exercising environmental rights and responsibilities among the diverse communities in the Arab Gulf are curtailed. Citizens typically receive significant state subsidies and their engagement with state authorities is meagre, giving them little incentive to rethink and alter consumption or opportunity to seek debate and action. Government initiatives aim to reframe subsidies and persuade citizens to pay for environmentalism based on an assumption of individual duty, treating them primarily as ideal consumers of environmental goods and services. The sponsor status of citizens obligates duties to non-citizen migrants while being prone to their exploitation and places responsibilities toward the extensive migrant populations on individual citizens themselves, creating a diffusion or obfuscation of responsibility. As the kafala system limits the rights of the migrant majority in the GCC, a major area of social and ecological relations goes overlooked. It is a social system of transnational labor that contributes to reproducing the environmental contradictions of citizenship. Thinking in terms of environmental citizenship helps to reveal how these ideologies of transnationalism and extractivism run deeply through and enable the specific privileges and consumerism of national citizenship in the GCC. National identity and national heritage work to naturalize this relationship between citizens and foreigners, and tend to elide or occlude extractivism and the migrant labor regime, mobilizing instead pre-oil heritage subjects, tourist destinations and consumerism. There is also a lack of contextualizing the system of labor, petro-industries and culture of consumption within their specific climatic and environmental settings in the region.

I suggest that the GCC configuration of labor, belonging and national identity, which is partly enabled through kafala, will be at odds with a version of citizenship that conceives environmental rights and duties as attached not only to citizens defined narrowly by circumscribed lineage but also to other people with significant relations to the state or society, even if these connections are the patron-client relations made possible by kafala over generations. As Dito (2015) suggests, redefining citizenship in GCC nations in order to overcome some of the adverse economic and social effects of kafala that have developed would imply a change to the financial privileges and patron status with which citizens are endowed, but bringing with it other benefits and challenges. We can add to this that steering citizenship (whether by drawing on Islamic traditions, private women's movements or other movements that produce social change) in a direction less impeding to a practical environmental ethics in relation to large groups of people (non-citizens) in the GCC would also carry the same implications.

As GCC states apply a framework of environmental citizenship that is strictly national, or transnational (even in its simple inter-territorial sense), they will likely fail to see that efforts or movements that can overcome barriers to action on environmental issues in the GCC might depend more on other, non-state relationships such as family ties in private domains, or cosmopolitan and labor relations within and beyond the stranger majority. More attention could be paid to the contradiction between, on the one hand, the deeply embedded history of Indian Ocean mobility and intermingling of people in the Arab Gulf and the hyper-dependence on migrants in GCC states today — both of which formed the past and present fabric of these societies — and on the other hand, an invisibility and lack of political understanding of emigres, migrants and others in constructions of the nation and environment or in pathways to fostering a deeper understanding and awareness of the ecological dimension of the citizen-foreigner relationship. A combined religious and secular consideration would also be of benefit in understanding frameworks of environmental citizenship in the Arab Gulf littoral, an approach that could attend to religious organizations, tribal traditions, and more private domains of life and associations where norms and practices emerge and are fostered with environmental and political significance, as much as it considers government policies that are espoused with the aura of state authority and officialdom.

## Acknowledgments

I would like to thank two anonymous reviewers, the guest editor of this special issue Benito Cao, and the commissioning editor of the journal Adela Alfonsi for their helpful comments.

## Disclosure statement

No potential conflict of interest was reported by the author.

## ORCID

*Ian R. Simpson* http://orcid.org/0000-0003-4480-8434

## References

Abdel-Raouf, M. A., & Luomi, M. (Eds.). (2016). *The green economy in the Gulf.* New York: Routledge.

Abdelraouf, M., & Azar, E. (Eds.). (2017). *Sustainability in the Gulf: Challenges and opportunities.* Abingdon, Oxon: Routledge.

Abuzinada, A. H., Barth, H.-J., Krupp, F., Böer, B., & Abdessalaam, T. Z. A. (Eds.). (2008). *Protecting the Gulf's marine ecosystems from pollution.* Basel: Birkhaeuser Verlag AG.

Akhonbay, H. M. (Ed.). (2018). *The economics of renewable energy in the Gulf.* Milton: Routledge.

Al-Rasheed, M. (2005). *Transnational connections and the Arab Gulf.* London: Routledge.

Al-Ubaydli, O. (2015). *The economics of migrant workers in the GCC.* Washington, DC: Arab Gulf States Institute in Washington.

Alam, T., & Luomi, M. (2018). *Engaging Gulf non-state and subnational actors in implementing the Paris agreement.* Abu Dhabi: Emirates Diplomatic Academy.

Bell, D. (2014). Environmental citizenship: Global, local and individual. In P. G. Harris (Ed.), *Routledge handbook of global environmental politics* (pp. 347–358). London: Routledge.

Benyāmin, D. (2016). *Goat Days.* (J. Koyippally, Trans.). London: Seagull Books.

Bristol-Rhys, J. (2012). Socio-spatial boundaries in Abu-Dhabi. In M. Kamrava, & Z. Babar (Eds.), *Migrant labor in the Persian Gulf* (pp. 59–84). London: Hurst.

Bryde, D., Mouzughi, Y., & Al Rasheed, T. (2015). *Sustainable development challenges in the Arab states of the Gulf.* Berlin: Gerlach Press.

Bulkeley, H., Andonova, L., Betsill, M. M., Compagnon, D., Hale, T., Hoffmann, M. J., ... Vandeveer, S. D. (2014). *Transnational climate change governance*. Cambridge: Cambridge University Press.

Cadène, P., & Dumortier, B. (2013). *Atlas of the Gulf states*. Leiden: Brill.

Campbell, B. (2014). Linking local and global in the sustainable development of biodiversity conservation. In P. Sillitoe (Ed.), *Sustainable development: An appraisal from the Gulf region* (1st ed., pp. 197–220). New York: Berghahn Books.

Chalcraft, J. T. (2010). *Monarchy, migration and hegemony in the Arabian Peninsula*. London: London School of Economics and Political Science.

Chanda, R., & Gupta, P. (2018). Indian migration to the Gulf: Overview of trends and policy initiatives by India. In P. Fargues, & N. M. Shah (Eds.), *Migration to the Gulf: Policies in Sending and Receiving countries* (pp. 179–197). Gulf Research Center: Cambridge.

DeSilvey, C. (2017). *Curated Decay: Heritage beyond saving*. Minneapolis: University of Minnesota Press.

Dito, M. (2015). *Kafala*: Foundations of migrant exclusion. In G. C. C. Labour Markets, A. A.-H. Khalaf, O. AlShehabi, A. Hanieh, & M. Patrick (Eds.), *Transit states: Labour, migration and citizenship in the Gulf* (pp. 79–100). London: Pluto Press.

Dobson, A. (2003). *Citizenship and the environment*. Oxford: Oxford University Press.

Dresch, P. (2005). Debates on marriage and nationality in the United Arab Emirates. In P. Dresch, & J. P. Piscatori (Eds.), *Monarchies and nations: Globalization and identity in the Arab states of the Gulf* (pp. 136–156). London: I. B. Tauris.

Dresch, P. (2006). Foriegn matter: The place of strangers in Gulf society. In J. W. Fox, N. Mourtada-Sabbah, & M. al-Mutawa (Eds.), *Globalization and the Gulf* (pp. 200–222). London: Routledge.

DubaiStatisticsCenter. (2018). *Number of Population Estimated by Nationality - Emirate of Dubai (2016–2018)*. Retrieved from https://www.dsc.gov.ae/Report/DSC_SYB_2018_0120_2003.pdf

EWS-WWF. (2015). *EWS-WWF Strategy 2015–2020*. Abu Dhabi: Emirates Wildlife Society-WWF.

EWS-WWF. (2018). *Enabling the UAE's energy transition: Top Ten Priority areas for renewable energy policymakers*. Abu Dhabi: Emirates Wildlife Society in association with WWF.

Fargues, P., & Shah, N. M. (Eds.). (2018). *Migration to the Gulf: Policies in Sending and Receiving countries*. Cambridge: Gulf Research Center.

Fattah, H. M. (1997). *The politics of regional trade in Iraq, Arabia, and the Gulf, 1745–1900*. Albany, NY: State University of New York Press.

Fox, J. W., Mourtada-Sabbah, N., & al-Mutawa, M. (2006). *Globalization and the Gulf*. London: Routledge.

Gardner, A. (2010). *City of strangers: Gulf migration and the Indian community in Bahrain*. Ithaca: ILR Press.

Gardner, A. (2012). Why do they keep comming? Labor migrants in the Persian Gulf states. In M. Kamrava, & Z. Babar (Eds.), *Migrant labor in the Persian Gulf* (pp. 41–58). London: Hurst.

Günel, G. (2019). *Spaceship in the desert: Energy, climate change, and urban design in Abu Dhabi*. Durham: Duke University Press.

Hailwood, S. (2005). Environmental citizenship as reasonable citizenship. *Environmental Politics, 14*(2), 195–210.

Harrison, R. (2015). Beyond "natural" and "cultural" heritage: Toward an ontological politics of heritage in the age of Anthropocene. *Heritage & Society, 8*(1), 24–42.

Harvey, D., & Perry, J. A. (Eds.). (2015). *The future of heritage as climates change: Loss, adaptation and creativity*. London: Routledge.

Hopper, M. (2015). *Slaves of one master: Globalization and slavery in Arabia in the age of empire*. New Haven: Yale University Press.

Horton, D. (2006). Demonstrating environmental citizenship? A study of everyday life among green activists. In A. Dobson, & D. Bell (Eds.), *Environmental citizenship* (pp. 127–150). Cambridge, Mass.: MIT Press.

ILO. (2019, June 29a). Labour Migration [Facts and figures]. Retrieved from https://www.ilo.org/beirut/areasofwork/labour-migration/lang–en/index.htm

ILO. (2019, June 29b). Labour Migration in the Arab States. Retrieved from https://www.ilo.org/beirut/areasofwork/labour-migration/WCMS_514910/lang–en/index.htm

Kapiszewski, A. (2001). *Nationals and expatriates: Population and labour dilemmas of the Gulf cooperation council states* (1st ed). Reading, England: Ithaca Press.

Khalaf, S. (2000). Poetics and politics of newly invented traditions in the Gulf: Camel racing in the United Arab Emirates. *Ethnology, 39*(3), 243–261.

Khan, N. Y., Munawar, M., & Price, A. (Eds.). (2002). *The Gulf ecosystem: Health and sustainability.* Leiden: Backhuys.

Krupp, F., & Abuzinada, A. H. (2008). Impact of oil pollution and increased sea surface temperatures on marine ecosystems and biota in the Gulf. In A. H. Abuzinada, H.-J. Barth, F. Krupp, B. Böer, & T. Z. A. Abdessalaam (Eds.), *Protecting the Gulf's marine ecosystems from pollution* (pp. 48–56). Basel: Birkhaeuser Verlag AG.

Latta, P. A. (2007). Locating democratic politics in ecological citizenship. *Environmental Politics, 16*(3), 377–393.

Longva, A. N. (1997). *Walls built on sand: Migration, exclusion, and society in Kuwait.* Boulder, Colo.: WestviewPress.

Luciani, G. (2014). The political economy of energy reform. In G. Luciani, & R. Ferroukhi (Eds.), *Political economy of energy reform: The clean energy-fossil fuel balance in the Gulf states* (pp. 7–24). Berlin: Gerlach Press.

Luciani, G., & Ferroukhi, R. (Eds.). (2014). *Political economy of energy reform: The clean energy-fossil fuel balance in the Gulf states.* Berlin: Gerlach Press.

Luomi, M. (2015). *The international relations of the green economy in the Gulf: Lessons from the UAE's state-led energy transition.* Oxford: Oxford Institute for Energy Studies.

Mahdavi, P. (2012). Informality and its discontents: Mapping migrant worker trajectories into Dubai's informal economy. In M. Kamrava, & Z. Babar (Eds.), *Migrant labor in the Persian Gulf* (pp. 85–103). London: Hurst.

Ministry of Climate Change and Environment, U. (2017). *National climate change plan of the United Arab Emirates, 2017–2050.* Dubai: Ministry of Climate Change and Environment, UAE. Retrieved from https://www.moccae.gov.ae/assets/30e58e2e/national-climate-change-plan-for-the-united-arab-emirates-2017-2050.aspx

National Center for Statistics and Information. (2018). *Statistical year book.*

National Center for Statistics and Information. (2019, June 29). National summary data page (NSDP)-Oman, demographic and selected socio-economic indicators [Population data]. Retrieved from https://apps1.ncsi.gov.om/nsdp/?1=131321

National Centre of Meteorology. (2019). *Climate yearly report 2003–2018, United Arab Emirates.* Retrieved from https://www.ncm.ae/en/climate-reports-yearly.html?id=26

NOAA National Centers for Environmental Information. (2018). *State of the climate: Global climate report for June* 2018. Retrieved from https://www.ncdc.noaa.gov/sotc/global/201806

Ouis, P. (2002). 'Greening the Emirates': The modern construction of nature in the United Arab Emirates. *Cultural Geographies, 9*(3), 334–347.

Pandya, A. A., & Sparding, K. (2016). Workers and the green economy in the Gulf cooperation council countries. In M. A. Abdel-Raouf, & M. Luomi (Eds.), *The green economy in the Gulf* (pp. 183–211). New York: Routledge.

Poirine, B. (2003). Managing the commons: An economic approach to pearl industry regulation. *Aquaculture Economics & Management, 7*(3), 179–193.

Robbins, P., & Sharp, J. T. (2003). Producing and consuming chemicals: The moral economy of the American Lawn. *Economic Geography, 79*(4), 425–451.

Seyfang, G. (2005). Shopping for sustainability: Can sustainable consumption promote ecological citizenship? *Environmental Politics, 14*(2), 290–306.

Sheppard, C., Al-Husiani, M., Al-Jamali, F., Al-Yamani, F., Baldwin, R., Bishop, J., … Zainal, K. (2010). The Gulf: A young sea in decline. *Marine Pollution Bulletin, 60*(1), 13–38.

Shriadah, M. A. (1999). Oil contamination along oil tanker routes off the United Arab Emirates (The Arabian Gulf and the Gulf of Oman). *Bulletin of Environmental Contamination and Toxicology, 63* (2), 203–210.

Sillitoe, P. (2014). *Sustainable development: An appraisal from the Gulf region* (1st ed.). New York: Berghahn Books.

Simpson, I. R. (2014). Concern amid the oysters as pearling is honoured: Nature and the environment in heritage practice. In K. Exell, & T. Rico (Eds.), *Cultural heritage in the Arabian Peninsula: Debates, discourses and practices* (pp. 31–47). Farnham, Surrey, England: Ashgate.

Stephenson, M. L., & Vyas, A. (2017). Deciphering the environmental impact of tourism development in Dubai: Developing sustainable directives and a citizenship agenda. In M. Stephenson, & A. Al-Hamarneh (Eds.), *International tourism development and the gulf cooperation council states: Challenges and opportunities* (pp. 204–221). London: Routledge.

Tsing, A. L. (2015). *The mushroom at the end of the world: On the possibility of life in capitalist ruins.* Princeton: Princeton University Press.

UAE Government. (2019, June 29). About the UAE [Fact sheet]. Retrieved from Retrieved from https://www.government.ae/en/about-the-uae/fact-sheet

Vivekanand, P. V. (2008). *Pearls of Dubai: Indo-Arab relations 1900–1958.* Dubai: Quality Printing Services.

Ward, C., & Ruckstuhl, S. (Eds.). (2017). *Water scarcity, climate change and conflict in the middle east: Securing livelihoods, building peace.* London: I.B. Tauris.

Willoughby, J. (2006). Ambivalent anxieties of the South Asian-Gulf Arab labor exchange. In J. W. Fox, N. Mourtada-Sabbah & M. al-Mutawa (Eds.), *Globalization and the Gulf* (pp. 223–243). London: Routledge.

World Bank. (2019, June 29). World development indicators [Databank]. Retrieved from https://databank.worldbank.org/reports.aspx?source=2&country=OMN,ARE,SAU

World Meteorological Organisation. (2019, June 29). World weather information service for Muscat (Seeb), World meteorological organisation [data]. Retrieved from http://worldweather.wmo.int/en/city.html?cityId=112

Yaghi, A., & Alibeli, M. (2017). Theoretical and empirical analysis of citizens' willingness to pay: Ethical and policy implications for the environment in the United Arab Emirates. *Public Integrity*, *19*(1), 41–57.

# Environmental citizenship for inclusive sustainable development: the case of Kelab Alami in Mukim Tanjung Kupang, Johor, Malaysia

Serina Rahman

**ABSTRACT**
Using environmental citizenship as the basis of its activities and initiatives, a community organisation, Kelab Alami, successfully combined environmental action with a socio-ecological systems approach to counter environmental and social injustice. The organisation fused traditional ecological knowledge with science for use in community-initiated ecotourism; engaged with surrounding developers, local authorities and the state government to mitigate environmental damage and maximize job opportunities for the community; nurtured and supported local entrepreneurship efforts and continues to spread environmental awareness and protection. This paper traces the development of Kelab Alami, the training and empowerment of youthful citizen scientists, and its attempt to garner multi-level support and collaboration for its environmental actions. This is a grounded illustration of resource mobilization in action, and a local view of a community's efforts to cope with poverty, environmental degradation and urbanization; a model that could be adapted and adopted by other communities facing similar circumstances and threats.

## Introduction

Island Southeast Asia is known to be a biodiversity hotspot, but many of these thriving tropical habitats are severely at risk from coastal development, urbanization, and pollution. Not only are the region's natural assets in trouble, but traditional livelihoods such as fishing and the cultural heritage that it encompasses are also highly endangered. Added to that are traditional social inequalities stemming from long-standing hierarchies, with absolute powers such as royalty, politicians and powerful businessmen at the top and marginalized local and indigenous communities at the bottom-most rungs of society.

The communities of the western Tebrau Strait between Singapore and Malaysia largely comprise Malay and Chinese fishermen who have either been displaced several times over history or targeted in ethnic massacres during the Japanese occupation. Some of the about 150 fishermen who remain continue to eke out a living at sea, but many have moved away or resigned themselves to other occupations as fish landings decline and natural areas decrease due to urbanization and coastal development (Rahman, 2019). Indigenous Orang Seletar who see themselves as originating from an island off Singapore also use these waterways; they too have been displaced from their marine territories to villages upriver.

This paper traces the development, successes, and challenges faced by Kelab Alami Mukim Tanjung Kupang (henceforth referred to as Kelab Alami), a community organization from Mukim Tanjung Kupang, Johor (Malaysia's southern-most state), a sub-district in this part of the Tebrau Strait, and how they evolved from an environmental education club to an organization of youthful environmental citizens working to counter social imbalances and environmental injustice. Using environmental education and citizen science as a base for their work, Kelab Alami now promotes community participation and benefit from the urbanization and development that threatens traditional livelihoods through ecotourism and other initiatives.

## Background to Kelab Alami and Mukim Tanjung Kupang

Kelab Alami began as an environmental education club for the youth of Mukim Tanjung Kupang in 2008. Prior to its establishment, I was invited by local teachers to help with the environmental education clubs of two local schools (Tanjung Kupang Government Primary School and Complex Sultan Abu Bakar Government Secondary School). This proved challenging, however, because of the limited time allocated for co-curricular activities and the schools' reluctance to allow students to experience natural habitats beyond school grounds. To overcome these difficulties and enable more youth to participate in its environmental and other programs, Shalan Jum'at (a Mukim Tanjung Kupang local) and I set up Kelab Alami. This allowed children from other schools (there are five in the sub-district) to attend weekend activities with their parents' approval. Mukim Tanjung Kupang comprises a string of fishing villages with many locals related to each other in a complex family network. Word of the club and its activities spread quickly throughout the sub-district and youth would turn up at Kelab Alami's rented clubhouse on foot, by bicycle or dropped off by their fathers on their way out to sea. I also often did a round of the sub-district by car to pick up those who had no means of getting there.

Although activities began in 2008, the club was only able to get state-government approval and official registration in November 2009, after some lobbying by a local elder, Arwah Pak Mus. I am not originally from this community, and since Kelab Alami was intended to be a locally driven initiative, I remained only an advisor and facilitator of the organization, and continued to conduct programs as required by the youth and community pro bono. The official organization structure is filled by local community members with some of the most important members being: Shalan Jum'at as founder and main decision-maker, Arwah Pak Mus as advisor and community patron, and Zurina Jaafar (the local teacher who initially invited me to the sub-district) as the head of the organization's local committee. While these committee members took responsibility for Kelab Alami, the spur for the club was my interest in teaching local youth about marine habitats. Serendipity linked me to locals who wanted to improve the lot, knowledge and behavior of the youth, and that kickstarted the community organisation. Local youth between the ages of 4 and 18 joined the organization as members and participated in activities ranging from farming to experiential science, estuary and marine habitat documentation and monitoring, ecotourism guide training, as well as team-building, public speaking and presentation skills, English classes and other community-centric activities such as graveyard and road cleaning.

While my initial intention was only to teach local youth about their natural habitats, Shalan's intention was to provide useful and productive activities for the youth, partly to distract them from drugs and other local problems, and to prove that young people can be engaged in positive activities without any need for religious or political involvement. In 2008, the community was facing some upheaval as rumors had surfaced of oncoming land seizures, resettlement and accelerated coastal development. While there are government-appointed leaders in the sub-district, as well as a regional fishing association, the community did not feel that they were trustworthy as they were believed to be negotiating with external entities for their own enrichment. Hence the community depended on individuals such as Arwah Pak Mus to help them in their daily difficulties and other needs.

A frequent visitor to the area at the time was an NGO that focused on the protection of an endangered marine species in the area, but had resorted to using (not engaging) the local community and opposition party politicians to achieve its goals. This NGO had initiated 'volunteer opportunities' for the general public (not the local community) to help in species documentation. At the same time, the NGO would provide participants with information on the local seagrass habitat. The costs paid by 'volunteers' would go to the NGO, with a portion of it paying (a highly discounted rate) for boat rental. The group's ability to find the species, however, depended on the knowledge of the boatmen, who knew where and how to spot them even as they camouflaged themselves amongst the seagrass. Yet little benefits of this 'tour' went to the community as the NGO's primary priority was its research and protection of the habitat from development for the sake of the endangered species of interest.

When asked separately about its involvement with the community, NGO members said that the community is more interested in money than protecting the environment. However, the NGO succeeded in manipulating the community into participating in a protest at the state government office. Liu, Wang, and Dang (2018) report on gaps between the goals of a post-disaster NGO and the local community in China; these differences in goals were also apparent in this situation. When asked by the NGO to participate in a marine blockade of a mangrove area, ostensibly to stop planned clearing of coastal mangroves, local fishermen refused to participate as it would prevent them from using the time and expense on petrol to catch fish. When their daily survival depended on the catch they make, they could not take the chance to participate in the blockade and not earn income for the day. When the NGO's members returned to their state of origin (the group did not originate from Johor), the local community had to endure the consequences of the local authorities' displeasure with the community's 'involvement' in the NGO's activities. These consequences included extraordinarily strict monitoring of road rules in the villages, obstructing the fishermen at sea and other hassles.

This scenario thus made it more difficult for the community to register Kelab Alami as an environmental club even though its intention was merely to provide better opportunities for local youth. Once it was finally registered, Arwah Pak Mus also had to protect it from appointed local and political leaders who invariably deemed it a threat to their authority and development plans. Some, more conservative, villagers expressed concerns about my presence in the villages (as a single city-born female) and the (potentially negative and wayward) influences on local youth that may arise from my involvement in the organization. They also actively tried to discourage youth from joining the program.

In spite of this, there were more than 40 youths participating in our activities at the time, and today there are 15 still actively involved and driving the organization's programs. Given the local resistance and suspicions in its early days, Kelab Alami had to find a way to function while trying to manoeuvre around local cultural biases and views (largely around the position and role of girls, women and youth, and the perceived 'impertinence' of their taking action to advise or teach men and elders about appropriate environmental behavior). There were also political complications of unintentionally gaining recognition for positive action and having that reflect badly on the local leadership because those actions were not of their initiation. In order to visibly demonstrate that they were engaged in positive activities, Shalan and the youth set up an organic vegetable farm behind the clubhouse (entirely by hand as they only had a grass-cutter, machetes, shovels and hoes). This helped to offset family household expenses as youth took home the harvests for their dinners. I then used the farm as a vehicle to begin teaching scientific principles, documentation and monitoring before introducing coastal and marine science.

As artisanal fishing families, most of the community live below the national poverty-line. Some women sell food in front of their homes or work as cleaners to supplement dwindling fishing incomes and young adults who did not become fishermen work as security guards or in factories. Becoming a lorry driver or getting a job in the nearby port garners local respect as it promises wages of at least RM2500 a month (about USD625). Local women are especially proud if their daughters marry immigration or customs officers (the sub-district lies exactly at the border between Malaysia and Singapore) as these government officers are deemed to be a step up and out of the villages.

In order to widen community support and strengthen community spirit, Shalan started a Kelab Alami Football Club (KA F.C.) for older youth (aged 19-28) and local women. The older youth (many of whom were unemployed and idle) were then roped in to supervise Kelab Alami members as they conducted their activities in the natural habitats. With some local women in support of Kelab Alami (those who joined the football team and their friends) and local fishermen in support of KA F.C., Kelab Alami youth had a ready audience to listen to their skits, songs, and talks at festive celebrations organized by Kelab Alami for the community. This locally created knowledge helped to spread environmental awareness across Mukim Tanjung Kupang

This study is an ethnographic reflection of how Kelab Alami evolved from just an environmental education club to an organization that uses environmental citizenship as the basis for engagement with businesses, state and government agencies and other external stakeholders whose decisions severely impact the community. The organization is of particular interest because of how a group of youth with little understanding of science, supported by a local champion, its founder, Shalan Jum'at, could evolve into an award-winning organization for the protection of the environment.

Although I was involved in the program from the start, and was instrumental in enabling the transformation of the youth, all of Kelab Alami's activities functioned on local goals, desires, and visions. Even as the required facilitation and training to achieve these objectives were provided, I remained detached in the observation of the youth and community and refrained from dictating the paths chosen. I took on the role of both participant observer and scholar-activist, but while I had a part to play in the initiation of the program, all other contributions and facilitation were at the behest of the community. A decade's worth of participant observation of changes in behavior in the organization's founder, committee, and youth participants as a result of the environmental content gained, community exchanges (amongst themselves) and interactions with external parties (other than myself) was collected. My role has mostly been to connect the community with external parties, translate views for both sides and become the vehicle through which the community's stories are shared. Copious ethnographic notes from informal interviews and conversations with both locals and external stakeholders are also drawn upon for this paper.

This paper brings together the fields of environmental citizenship with that of environmental education and citizen science (elaborated on in the following sections) and puts forward the story of a community that has evolved from passive recipients of urbanization and change to active proponents of community participation in development, so as to ensure that the wider community is not side-lined or deprived of access to the benefits of this change. This is resource mobilization (as outlined by Edwards & McCarthy, 2004) in action and the illustrations that follow are grounded accounts of how a community used both local and external networks and resources to enhance their ability to negotiate both within and beyond the village, sourced for and managed funds and markets for the outcomes of their community-driven research, and created knowledge that combined both scientific and local ecological knowledge for public and community consumption, as well as for state and national use and preservation. Trejos, Huang, and Chiang (2008) discuss the issues surrounding external NGOs working in a community, and how they might use 'tourism' to alleviate poverty in rural areas. Banks and Hulme (2012) and Islam and Morgan (2012) describe the difference between NGO and community-based organizations' interests and the former's inability to ascend beyond the lowest rungs of participation; this community's experience with the marine NGO proved that the best interests of the community are often not at the heart of external NGO action. Kelab Alami is an example of a community-driven group that is able to overcome such issues (Rahman, 2016), ensuring that community goals of alleviating poverty while ensuring sustainability, enhancing community capabilities and demonstrating that local knowledge is on par as that of 'certified' scientific knowledge is the priority of the programs conducted. This paper will contribute to a gap in the literature (Fan, 2008) by providing a grounded perspective of local efforts for change, its impacts and responses.

## Environmental education as the basis of environmental citizenship

In their discussion of world views of the environment, Altman and Chemers (1980) commented that Asians traditionally see themselves as part of nature, with both mutually responsible for each other's well-being. My documentation of local folklore and cultural history showed that this rings true when elders relate stories of their understanding of local weather systems and tides. They often talked about how in the past, they knew to respect the sea by guarding their language and behavior while on the water and taking steps not to offend spiritual and faunal inhabitants in return for the harvests it provided. Today this syncretic relationship with nature has subsided with the advent of Islam and increased religious orthodoxy, but the community is deeply affected by externally induced damage to their natural environments.

Artisanal fishermen in Malaysia are amongst the poorest populations in the country, with income levels and livelihood instability that would render them under the category of 'chronic poverty'; theirs is a 'multi-dimensional poverty of long duration … that is passed from one generation to the next' (Scott, 2006, p. 2). Those suffering this level of poverty are often highly dependent upon natural habitats for their livelihoods, and when there are shocks to the status quo (such as development and destruction of those habitats), they tend to add more pressure to the very habitats they depend on in order to survive. Interviews with fishermen and local elders revealed how they felt that they were now at the mercy of nature and its climatic changes instead of being an active part of it, and are helpless at the hands of those in absolute power (businessmen, politicians, and royalty) whose decisions result in irreversible damage to the natural ecosystems they depend on; especially when decisions are made either in spite of their presence or purportedly 'on their behalf.'

Drastic changes to this community began in the 1940s with the Japanese Occupation of Malaya and Operation Sook Ching (Ismail, Shaw, & Ooi, 2009), when many ethnic Chinese were massacred and daily routines were disrupted by the hazards of war. With independence and industrialization came the building of the Port of Tanjung Pelepas in 1997; the first of many coastal developments that forced the community to be resettled (Renkema & Kinlan, 2000). Although Kelab Alami was set up against this backdrop, Shalan's goal was not to protest these developments, but to help local youth cope with change. I, on the other hand, simply wanted the youth to understand and appreciate the area's unique intertidal seagrass meadows (the largest in Peninsular Malaysia), the mangroves, rocky intertidal shores and soft coral reefs of nearby Merambong Island, and the richly biodiverse coastal mangroves that stretch along the Malaysian coast of the western Tebrau Strait, and inland along both banks of the Pendas and Pulai Rivers.

Schild (2016) notes that environmental education is the basis of environmental citizenship; a means with which a community can take back control through realizing the impact of their actions to counter the helplessness that results from irrevocable development decisions by external entities and higher powers. Kelab Alami's environmental education program began with a series of outdoor experiential science classes anchored on the organic farm where youth learned skills such as observation, monitoring, and documentation of farm ecosystems, which were then transferred to nearby mangrove, seagrass and island habitats. These ecosystems were not unfamiliar to the youth, but provided them with the scientific skills to explore their intricacies and generated a deeper understanding and appreciation for areas that had previously been taken for granted. Sessions incorporated both theoretical and practical activities, with a lot of emphasis on encouraging the youth to speak and present information on their natural heritage to random strangers.

Kelab Alami's focus on natural areas familiar to the youth provided local context and leveraged on existing traditional ecological knowledge such as their understanding of fish behavior in relation to tides and wind. The objective at the time was only to provide scientific structure and methodology to content that they already knew. The youth were also shown how this knowledge could be converted into action, beginning with learning how to document and monitor the environment, to the creation of their own environmental education content and fora to share information with the rest of the community during festive celebrations.

Science education in rural Malaysia is not ideal. Teachers in rural schools often do not have the financial or administrative support to take lessons out of the classroom and local youth are often disinterested in classroom sessions or deemed academically incapable. My immersion in the community revealed, however, that there is a wealth of traditional ecological knowledge such as medicinal uses of local flora and fauna, animal behavior, and weather patterns that are not given due recognition. Dhillon (2017) discusses a parallel scenario where local knowledge is given scientific structure through citizen science in a Californian community's efforts to contest the siting of a waste facility. At this stage of Kelab Alami's existence, the goal was only to enhance local recognition of their own knowledge and engender self-respect to counter the continuous stream of NGOs and researchers who use them as informants without recognizing the community for the value of their contribution.

With little science background when they joined the club, the youths' expectations of the program was merely to have fun activities every weekend, perhaps learn something interesting, but also to enjoy food and drinks throughout the session and bring home farm crops for the family. I conducted this part of the program (except the maintenance of the farm), with activities ranging from science to arts and craft, games and ad-hoc public presentations centered on local marine habitat information. The youths' assignments to disseminate information to others in the community was the beginning of collective environmental action and the spread of wider environmental literacy by the youth – based entirely on what they deemed to be important to share with their friends and family (as opposed to an intervention by an outsider that dictated the content to be disseminated).

## Citizen science as an educational tool

As Kelab Alami gained community support, I broadened the program to include citizen science tools such as weather monitoring, flora and fauna observation skills, and habitat and species documentation and identification to enhance activities and increase the youths' self-confidence in their knowledge. Citizen science refers to the involvement of laymen in scientific exploration, activities and data collection. This can range from nature hobbyists to random adult or youth volunteers with no prior scientific experience. While there has been some debate over the validity of data collected by citizen scientists (Burgess et al., 2017; Riesch & Potter, 2014), some studies have shown that with proper structure, training and methodology, citizen science can be a reliable source of scientific information, especially in an area like Mukim Tanjung Kupang, where there is limited to no baseline information and longitudinal observations such as local ecological knowledge are necessary (Gregory & Atkins, 2018; Parrish et al., 2018; Theobald et al., 2015).

There are three levels of citizen science (Grace-McCaskey, Iatarola, Manda, & Etheridge, 2019): at the lowest level are contributory projects where volunteers provide information based on requirements dictated by professional scientists. Beyond that is collaborative citizen science where participants also develop project goals and analyze data among other tasks. Co-created projects are the ideal, but are rare; where all aspects are co-designed between the local community and professional scientists, and the research and dissemination of outputs are driven by the local community. At this initial stage (early 2010), I used contributory citizen science such as SeagrassWatch monitoring methodologies as a tool to enhance seagrass habitat understanding. This tool also enabled the community to gather baseline data that is still used to compare changes in habitat health and species presence. Similar monitoring approaches were created or adapted from other scientific monitoring protocols to strengthen mangrove and rocky shore habitat learning.

There is unanimous agreement amongst experts on the value of citizen science in enhancing science education and natural habitat appreciation (Chandler at al, 2017; Cigliano et al., 2015; Dean, Church, Loder, Fielding, & Wilson, 2018). It is also seen as an inexpensive way to both gather data and increase environmental awareness (Dickinson et al., 2012). Citizen science programs are reported to democratize science; allowing the inclusion of marginalized or minority groups, enabling them to participate in science instead of being mere recipients of science, as well as provide opportunities for local stewardship and investigation into issues that matter locally (Bonney et al., 2014; Dickinson et al., 2012; Pandya, 2012). Kelab Alami was set up and initially financed by Shalan's and my personal funds. As the program expanded and achieved attention (when the youth began to offer guided tours) it was able to garner small grants from the Shell Sustainable Development Grants program and the Ford Foundation (between 2010 and 2013). This funding demonstrated recognition for the organization and the youths' ability to conduct credible science at the local level.

A network of scientific advisors and nature experts (my contacts and friends) supported the program by volunteering to visit to teach the youth, or guiding them through *in-situ* experiments that enabled them to become more familiar with the science behind their habitats. While the youth had vast local knowledge, it was also important to widen their horizons; when funding was available, they were taken on field trips with these scientist-advisors to other parts of Malaysia, at times presenting scientific lectures to university students in Kuala Lumpur and hosting workshops with residents of other island communities about shared megafauna such as dugongs and turtles. Altman and Chemers (1980) note that environmental attitudes match the extent of an area travelled; these excursions enabled the youth to apply their knowledge elsewhere, strengthen their ability to teach others about the environment, and broaden their attitudes and action for the habitats that they had become ambassadors of. This empowered them by allowing them to be the voice of their community. It also increased their social standing in the eyes of the community as they were able to use the knowledge gained from the collection of scientific data to represent the community at events in the national capital, Kuala Lumpur, as well as earn extra pocket money as shown in the following section.

The sharing of knowledge beyond their village boundaries was the youths' early steps in environmental citizenship. As they taught others about their natural habitats, they also strengthened their learning of environmental and conservation science (Gebbels, Evans, & Delany, 2011) and paved the way for more transformational aspects of the program. It was vital at this stage that the youth understood that the knowledge gained from the program was to be shared, and that they took ownership of the content and messaging that was extended to the wider community. Just as Iverson and Jóndóttir (2019) highlighted the importance of democratic participation, the youth in Kelab Alami were shown that they could and should take pride in their natural habitats and improve their understanding of the ecosystems so as to be able to share their treasures with others.

This sharing went beyond the immediate community and field trips; the youth were trained as ecotourism guides and were thus tasked with crafting their own tourism content based on an amalgamation of local ecological knowledge, folklore, heritage stories, and science. Their training as citizen scientists and habitat monitors helped them to guide with scientific legitimacy and their familiarity with their natural playgrounds meant a wealth of generational information that no outsider could teach them. For example, while they were able to describe the biological cycle of a turtle based on the science they learnt in the program, they also added on local knowledge such as locations of frequent turtle sightings, oft-observed turtle behavior and local beliefs and mythology related to turtles and their presence. They were taught that theirs was a duty to 'enhance sustainability for the collective' as mentioned by Iverson and Jóndóttir (2019); the first step towards environmental citizenship and responsible action for the wider society.

Most of the 'tourists' who joined the guided habitat walks at this stage were groups from Kuala Lumpur or Singapore arranged through friends, but at times these widened into company excursions and school visits. There was an economic incentive behind the tourism effort. In the early stages visits took place bi-monthly; the youth were able to earn an allowance (initially RM10-RM20 per session (USD2-5) plus tips), but today they earn RM50-RM100 per visit (USD12-24 depending on the scope of the work required) from the fees paid and the fishermen earn extra incomes as boatmen (ranging between RM100-300 (USD24-72) depending on whether they had their own boat and petrol) for a 3–4 hour visit. An awareness of the fiscal benefit that a natural environment can bring is an 'incentive for the greater good' (Dobson, 2007) and reinforces the conservation narrative the youth shared with their community. This was the beginning of the effort to counter environmental and social injustice.

## Environmental citizenship for social justice

By definition, sustainable development requires environmental stewardship and protection (Clarke & Agyeman, 2011). In 2014, a number of large-scale mixed developments began to pick up speed in the western Tebrau Strait, on both sides of the Malaysia-Singapore border. Several of these coastal projects included the reclamation of new islands comprising residential and business districts, petrochemical storage facilities, shipyard and port extensions, as well as extensive development within mangrove and riverine systems. While many of these projects advertised their efforts at sustainability, this did not include similar concerns for local livelihoods; fishermen were suffering from drastically declining fish catch and the sea, in their words, 'was getting smaller.' They had to spend more on petrol costs, as well as take higher risks in their small fishing boats to travel further out to the Melaka Strait, at times encroaching into Singapore or Indonesian waters to ensure that they could bring home sufficient catch.

It is widely recognized that environmental protection is essential for poverty alleviation (James, 2019), especially when rural or traditional livelihoods are dependent on the health of natural habitats. Their desperation to put food on the table and bring home sufficient earnings leads to more resource extraction and a disregard for syncretic practices of old, where only what is immediately needed is taken from the sea. With future catches no longer guaranteed, there was now a tendency to over-harvest, taking in juvenile species that would, in the past, have been tossed back to sea. This is a vicious cycle. Externally initiated damage to nearshore habitats because of development that disregards local livelihoods (exacerbated by the effects of climate change) leads to locally intensified extraction in the surrounding waters. The solution to environmental issues requires that social injustices are addressed (Smith & Pangsapa, 2008). In order to break this cycle of destruction, community livelihoods need to be saved or enhanced.

It was at this stage (2014) that Kelab Alami moved into a second phase of evolution. From being a mere environmental education club, Shalan Jum'at decided to evolve its goal into one of enabling the community to participate in and benefit from development. The organization began to reach out to local agencies, such as the Iskandar Regional Development Authority (IRDA), for advice. IRDA pledged to support the community through further environmental engagement and community development. At the same time, Kelab Alami was approached by the biggest developer in the area, Country Gardens Pacificview Sdn Bhd, with an offer to support the continuation of the organization as part of its contribution to the villages surrounding its property, and given the youths' exhibited expertise on local habitats and their efforts for the community. This spawned the first of several engagements with other developers such as Sunway Iskandar and Iskandar Investment Berhad and local authorities such as the Johor Port Authority and Johor Tourism, which helped the youth to enhance their tourism offering (still based on local habitats and fishing heritage) and broaden opportunities for supplementary incomes. This was one way in which the organization used its environmental knowledge to act against environmental injustice and poverty. The very developers who were damaging the coastlines were encouraged to act to mitigate their damage to the environment and support alternative livelihoods by funding some of the costs of tourism development and training. This helped to partially offset the negative impacts of their development projects.

The effort to counter social and environmental injustice also expanded in several other ways. Shalan set up a market to take in fish from the local fishermen. With financial support from Country Garden Pacificview Sdn Bhd, he paid more than the usual amount for the catch to the fishermen (monies which had previously gone to the middleman), yet maintained the usual prices to the customer. This meant that the fishermen doubled their incomes and had less need to over-exploit the surrounding waters. Shalan also encouraged the fishermen to stay within their nearshore waters to reduce their personal risks and costs, and return catch that had not yet reached breeding age. This was done through a combination of persuasion, peer pressure from the fishermen who already understood the need to prolong the health of their marine ecosystem, and financial coercion in terms of the minimal monies paid or rejection of unsustainable catch. Shalan also began to teach customers that fair compensation needed to be made to the fishermen for their effort and costs in bringing in the seafood. At the same time, he mobilized the community to begin the process of attaining passenger boat licenses for tourism, as well as to get involved in other efforts to preserve their fishing heritage and myriad other opportunities for supplementary incomes.

While the initial capital for this effort came from the developer's grant, Shalan was able to generate enough sales from the growing pool of fishermen who joined the initiative and a wide customer base to sustain the business. Customers found out about his market through word-of-mouth and some social media publicity. They included individuals from Johor and nearby Singapore, as well as restaurant owners and caterers from the nearby development projects and Johor Bahru.

With IRDA's support and training, Kelab Alami set up a cooperative to strengthen the benefits gained from tourism for the wider community. Cooperative members (restricted by invitation only to those living in Mukim Tanjung Kupang and unconnected to political elites) were able to provide ideas to improve tourism efforts, as well as support tourism through catering and homestays, which widened the opportunities for local women to get involved through heritage cuisine-related tourism. In 2010 when tourism began, there were no homestays in the village. Today there are at least five licensed homestay options, with one extensive cabin set-up especially for short stay rentals. IRDA also facilitated introductions to other agencies such as Johor Tourism and the Johor Department of Education, which then led to the licensing and recognition of the Kelab Alami youth as nationally recognized nature guides, as well as natural habitat experts involved in a new outdoor education curriculum by the state. In 2016, Kelab Alami was awarded the Iskandar Malaysia Social Hero (Organization) Award for Environmental Protection. Awangku Hassanal (2009) highlighted examples of other community groups in Sabah, East Malaysia that have also used ecotourism as a means to alleviate socio-economic difficulties; he cites the community's ability to meet its goals and self-empowerment as an indication of its success. Given the above achievements, it can be said that Kelab Alami too, is a success. However, the true measure of the community's empowerment only comes with recognition by others (in this case, state agencies) and their willingness to fund and support the community in its work. This is described in the following sections.

I worked with IRDA's environmental department to devise PESISIR, a coastal documentation, and monitoring program that intended to engage with multiple stakeholders, including local community, non-governmental organizations, researchers, businesses, and government agencies. Kelab Alami's existing network of scientist-advisors, its existence as a community organization and recognition by several state agencies and developers meant that it was an easy fit as the pioneering partner of this initiative. This then opened the door for Kelab Alami to engage with the Johor Port Authority, who agreed to fund research on dugongs, otters and seahorses.

## Environmental citizenship for agency and transformation

The support from the port authority for community-led research was another step towards effective transformation. Iverson and Jóndóttir (2019) note that for environmental citizenship to be truly effective, it must influence the decision-making process. Even before the second phase of Kelab Alami kicked in, the youth were called upon by their parents and other members of the community for information to be used in discussions with developers pertaining to the location of new development projects. While the youth were not directly involved given their age, their knowledge of the local habitats was seen as vital in negotiations with developers and the information was used to dissuade them from destroying vital habitats. Before deciding to fund Kelab Alami in 2015, Country Garden Pacificview Sdn Bhd asked for a consultation with Kelab Alami youth to learn about the natural ecosystems within their project site as they had been told that the youth were local habitat experts. It was this engagement that convinced the developer to support Kelab Alami, as well as to demonstrate their commitment to and concern for the local community and environment.

In their study of volunteer scientists' knowledge being incorporated into policy-making in the United Kingdom, Ellis and Waterton (2004) noted that while information by volunteer naturalists and lay citizens were vital in biodiversity action planning, some knowledge was deemed more valuable than others. This inequality was also experienced in Mukim Tanjung Kupang.

The area's recognition as a biodiversity hotspot meant that many scientists descended on it for research. Many tapped on local knowledge to find out where species were hidden, others hired locals as informants or boatmen; Kelab Alami youth were often hired as field research assistants by local and foreign universities because of their familiarity with scientific protocols and understanding of local species. However, most of these researchers did not return or share any of their findings with the community nor did they acknowledge the youth in their publications, even when the academics' 'discoveries' were contingent upon the locals to find what they needed. When I argued for fair treatment of the locals, or suggested collaborating on publications, the response was often that they (as the superior entity in terms of knowledge and academic qualifications) would publish something and the community could 'use it as a reference.' Very few scientists recognized the value or importance of local knowledge nor acknowledged the ability of the community to contribute to and conduct credible research and science.

The support by the port authority to fund Kelab Alami's research was a shot of encouragement for the organization and a vote of confidence for the youths' abilities. Study objectives and methodologies were devised by Kelab Alami youth and then supportive scientist-advisors from the University of Malaya and the National University of Singapore were consulted to verify that the approaches were sound. At times, these scientists took the youth into their labs to teach them specific skills and techniques such as seagrass and sedimentation analysis and GIS mapping, which were then replicated or applied in the field. Study findings were analyzed by the youth, then again run by the scientists to verify and validate. These and other studies were presented as papers and posters at international conferences such as the World Seagrass Conference 2018, held in Singapore, by the youth themselves. Environmental education and citizenship had thus evolved into effective and successful community science. Books (such as the *Living Heritage of Mukim Tanjung Kupang*, 2019), newsletters and posters sponsored by myriad developers have helped to offset knowledge inequalities and prove that community knowledge of the environment is on par with those of 'certified' researchers and not 'subservient.' This is citizen science at the highest level: where knowledge is co-generated and the community takes complete ownership and leadership of the research and its outcomes to meet their needs (Grace-McCaskey et al., 2019; Gregory & Atkins, 2018).

With this recognition of their abilities, research output and other information generated by Kelab Alami youth such as coastal habitat maps and species lists have been used by developers and local authorities. The Johor Port Authority, for example, has specifically asked Kelab Alami to advise them on coastal mangrove areas that should be protected and left untouched during port expansion works. Community observations of construction workers' behavior and action have been tapped in the mitigation of environmental damage by developers, and has also been used as a check and balance between developers and contractors, as well as between port authority and port. State government agencies have also tapped on Kelab Alami's experience in community engagement, empowerment for environmental citizenship and community ecotourism to advise on the initiation of similar initiatives in other locations. Kelab Alami is frequently approached to either expand or replicate its programs in other parts of Johor, such as Pontian and Sungai Melayu, which are also high biodiversity areas facing the threat of urbanization. Not only does this indicate that the environmental citizenship of these youth has influenced decision-making, but that in their struggle for land and livelihoods, they have offset some of the injustice that marginalized them (James, 2019). Recognition of their environmental knowledge and action has earned them political capital and agency to influence those traditionally seen to be above their social standing.

This type of external recognition has earned Kelab Alami political mileage within the community as shown in the following illustration. Village communities across Malaysia are steeped in traditional hierarchies. Appointed village and district leaders, political branch heads and fishing or farming association heads wield incontestable power and can either provide or block access by the average villager to government assistance, developer compensation and other benefits. When Kelab Alami was first set up, this local hierarchy had to be managed by local elders such as Arwah Pak Mus (who otherwise held no official position), thereby protecting the organization's effort to mobilize the youth and wider community. Kelab Alami actively worked around this hierarchy to ensure that the average villager had equal access to jobs and other opportunities (that were usually reserved for those with political or family connections to this middle layer of authority) by negotiating directly with developers for job opportunities for the villagers, then going door-to-door in the sub-district to inform residents of openings available. In the past, the wider community would not be allowed to apply for these positions (they were blocked by the local hierarchy) or would not even be aware that there were jobs available.

With increased visibility and clear connections to more powerful entities beyond the village such as developers and government authorities (all of which essentially stemmed from the youth's actions for the environment), attempts by the local hierarchy to monopolize job opportunities and maximize personal benefits from construction projects were reduced. These local leaders or political representatives were also unable to shut down the organization that they initially felt was a threat to their position and power given Kelab Alami's access to entities beyond the sub-district (a privilege previously only accorded to these superior members of the local hierarchy). This ability to have direct access to developers, state agencies and government, as well as enable access for others in the community is an illustration of Kelab Alami's political capital. The inability of those in the village hierarchy to block that access is an indication of success in combatting environmental and social injustices.

Armitage et al. (2017) noted that small coastal communities with a strong attachment to place and a direct connection between coastal resources and livelihoods can organize themselves to make decisions and instigate transformational change. Kelab Alami has embodied this integration of socio-ecological systems and environmental citizenship in response to environmental injustice.

Based on Armitage et al's socio-ecological systems framework (2017), the driver for Kelab Alami's ground-up action was the externally induced loss of natural environments and subsequently, the decline of traditional livelihoods. Over a decade, those involved in Kelab Alami evolved to strengthen their environmental knowledge, citizenship, and action. In their quest to document their natural and cultural heritage, the youths' sense of belonging to the location and ownership of the habitats strengthened. There was a clear increase in their determination to gather local knowledge before it disappeared with the passing of elders. There was also more interest in finding out more about their local and ethnic origins. This change in attitudes is similar to that reported by Fan (2008) as local activists in Taiwan worked to protect their township from the siting of a waste facility.

This evolution not only occurred amongst the youth, but also in Shalan Jum'at, who found his voice during the second phase of the organization and began to take complete ownership over the direction of the organization and the actions taken for the wider community. Even though the emphasis shifted to community capacity-building and support, all action was rooted in environmental protection and stewardship. Habitat documentation, research, and education continued with a wider audience, driven by the youth who were the initial recipients of the program and this environmental expertise and action continued to be the basis of the respect they earned from external entities. Transformation occurred as they built resilience to increasing and incessant depletion of their natural resources (by having supplementary sources of income in the face of deteriorating fish catch and fishing incomes) and changes to their neighbourhood; providing input into broad-ranging higher-level decisions by the local authorities or developers as described earlier, even as they sourced for entrepreneurship and supplementary income opportunities (through ecotourism and job placements in the new developments) for others in the community.

As the youth got older, those who could not study beyond high school (because they were not academically inclined) now work full-time for Kelab Alami and draw salaries from either funding or the founders' own incomes. Those who have continued to higher studies draw allowances for working part-time as guides or administrative assistants during their free time and school holidays. When asked why they opted to stay and work for Kelab Alami when they could have looked for jobs beyond the village, the youth responded that they wanted to work for the community and help their friends and family benefit from the change happening around them. This is indicative of the behavior of environmental citizens who have a 'commitment to the common good' as described by Dobson (2007, p. 280). Kelab Alami leveraged on multiple stakeholders, sponsors, volunteers, and experts of various levels; building vertical and horizontal networks to support the work for the community and create innovative partnerships to begin to counter environmental and social injustices such as habitat damage as a result of external and distant decision-making by those in power and the blocking of access and opportunities by those in the local village hierarchy.

## Local limitations and future struggles

While Kelab Alami has shown tremendous progress over the decade since its inception, its success is hampered by the extent to which its current abilities and recognition by government agencies and businesses can reach. While it is now able to bypass those in power at the local level and negotiate with developers and agencies on decisions that affect the community, it is not able to forcefully hold these entities accountable for their actions. Kelab Alami's success still depends on their willingness to take into account local views and needs. These far more powerful businesses and government agencies could easily withdraw their support and agreements with the community and pursue their goals with scant regard for their impacts.

Even though Malaysia has rules in place to protect the environment and provide compensation or assistance for the local community (to a lesser scale), enforcement is almost non-existent and there are many loopholes that those in power can use to avoid these requirements. The local fishing association, for example, is particularly belligerent in its selectivity in the distribution of allowances and livelihood compensation. Local fishermen constantly complain about irregularities and ambiguity in the distribution of subsidies and government assistance. There have been several efforts by the fishermen to dislodge those in power at the association and query murky accounting records, but the appointed leaders have thus far evaded all demands for accountability and transparency. Instead, they are able to use their connections with politicians and royalty to maintain control and even further abuse the fishermen by demanding unnecessary payments in return for jetty use and fishing license renewals. Even with a new federal government in place (as a result of the recent Malaysian general elections), the fishermen and Kelab Alami have not yet been able to (nor know how to) initiate charges against the association; instead, choosing to focus first on livelihood improvements to ensure the community's survival in the current difficult economic conditions.

In their explanation of the socio-ecological systems that local communities work within to protect their coastal commons, Armitage et al. (2017) note that there is always an unequal distribution of power, and a community's success at protecting their resources depends on their ability to mobilize to create conditions of justice and equity, leveraging on multiple networks to contain negative action. In the case of Mukim Tanjung Kupang, where traditional hierarchies are deeply entrenched, there is little hope of completely effecting change, but pressure can be added to keep some on their toes. Kelab Alami now functions under the total understanding (and some expectation) that all the habitats will be lost, that the villages may be lost completely, and that Kelab Alami could be shut down at any time by those in power. In spite of these very real possibilities, Kelab Alami has chosen to persevere for now.

Clarke and Agyeman (2011) cite the need to empower communities to be able to participate in decision-making processes, in the same vein as Local Agenda 21, and through the Just Sustainability Paradigm. These approaches prioritize social inclusion, equity, and justice in the delivery of sustainable development. This might be possible in urban areas where residents are seen to be credible and capable contributors to discussions, but amongst marginalized rural communities that are othered as inferior and uneducated, there is little expectation of their inclusion in these discussions. Those who are invited to any kind of negotiation are usually those appointed into positions of power (the local hierarchy such as village and fishing association appointed leaders), who often make decisions counter to the desires and needs of the people they purport to represent. The ability of the Kelab Alami youth to counter this perception is part of their attraction: they wildly supersede the sub-standard perception and expectations of 'mere village youth' and instead can hold discussions expected only of those with 'academic qualifications.'

It has been said that the use of environmental citizenship in the pursuit of sustainable development is an easy way out for the institutions that should be held responsible for the protection of coastal commons (Dimick, 2015), instead putting the weight of responsibility and civil action on the shoulders of individuals. Others point out that environmental citizenship is a far too simple answer to highly complex problems of natural resource preservation, livelihood issues, and inequality (Karlsson, 2012). James (2019), however emphasizes the need for collective action, as in the case of Kelab Alami, which has somewhat successfully brought together myriad local youth, fishermen and cooperative members to challenge more powerful actors – no matter the outcome. As recommended by James, Kelab Alami's attempts to alleviate poverty and institute environmental and social justice are multi-targeted, with numerous initiatives that are locally organized and implanted; all of which are specifically engineered to solve specific local problems. Mere environmental education is not enough to instigate change if citizen action does not exist.

Smith and Pangsapa (2008) stress that strategies for change should be based on the community's daily realities and their relationship with the environment. Ecological citizenship and action for local livelihoods need to be grounded in a local understanding of the need to transform and adapt approaches that work for them. This lends itself to the original aims of Kelab Alami as an environmental club: to facilitate and enhance local ecological knowledge, as well as to ensure democratic participation in the youths' creation of new knowledge based on the combination of science and traditional knowledge to be disseminated by the youth themselves to the wider community (and visitors) for knowledge and habitat preservation. Today, in line with its mention in the Johor Sustainability Policy 2017–2021 (UPEN, 2016, p.70), it is negotiating with surrounding developers to begin a locally led collaborative business-community habitat management program. The hope is that Kelab Alami will be able to build on their successes for the good of the people and environment of Mukim Tanjung Kupang.

## Conclusion

Kelab Alami Mukim Tanjung Kupang has demonstrated how environmental education can morph into environmental citizenship to respond to and neutralize environmental and social injustice for a community whose traditional livelihoods are threatened by externally induced causes. In spite of the success and progress that the organization has garnered over its decade of existence, the political capital it has gained is only effective within the constraints of traditional hierarchies; it may never be able to overcome the absolute powers (politically connected businessmen, politicians and royalty) that control myriad state-wide matters, should they suddenly decide to act against the community's interests.

In spite of the challenges, and the knowledge that the organization, natural habitats, and community as a whole may not have longevity at the hands of these aforementioned powers, Kelab Alami continues to pursue its goal to alleviate the difficulties of its community by providing alternative income options in the face of deteriorating traditional livelihoods and urbanization. The organization also works to protect what remains of its local ecosystems within hierarchical (and feudal) constraints by working with developers to mitigate damage and protect vital and sensitive natural areas. Today the organization is working to set up a nature and heritage center to showcase the outcome of their decade-long documentation and research, as well as function as a commercial center for their community. Even as they have earned accolades from external entities, they continue to have to prove themselves within the community as they attempt to instigate individual action for the environment, such as to improve waste disposal practices and change local attitudes towards waste and rubbish disposal through district-wide cleanliness campaigns.

It is clear that environmental citizenship has been a vital tool in the progress and success of Kelab Alami, as well as the basis for the acknowledgement and recognition that they have gained with important external stakeholders. However, for complete livelihood resurrection and sustainable development, several other components and initiatives have to exist at the same time. It is a model that can be adapted and adopted in other communities facing the same threats to their habitats and livelihoods, but as shown in this case, it would have to have a local champion such as Shalan Jum'at to push it through, and to meet local needs and realities.

## Acknowledgements

I would like to thank the editors, Benito Cao and Adela Alfonsi, and the anonymous reviewers for the invaluable comments and suggestions that greatly strengthened this paper. I am also immensely grateful to Shalan Jum'at, Zurina Jaafar, the late Arwah Pak Mus and the youth and community of Mukim Tanjung Kupang for allowing me to live amongst them and observe this evolution of environmental citizenship over the last 12 years.

## Disclosure statement

The author is the co-founder of Kelab Alami, driving the environmental education component in its initial phase. She subsequently remained in the background to facilitate and assist the community with external stakeholder engagement as they took ownership of the organization and moved it forward from 2012 onwards.

## References

Altman, I., & Chemers, M. (1980). *Culture and environment*. Monterey, CA: Brooks/Cole Publishing Company.

Armitage, D., Alexander, S., Andrachuk, M., Berdej, S., Brown, S., Nayak, P., ... Rathwell, K. (2017). Communities multi-level networks and governance transformations in the coastal commons. In D. C. A. Armitrage, & F. Berkes (Eds.), *Governing the coastal commons* (pp. 231–251). London: Earthscan from Routledge.

Awangku Hassanal Bahar Pengiran Bagul. (2009). *Success of ecotourism sites and local community participation in Sabah*. Unpublished PhD thesis submitted to Victoria University of Wellington.

Banks, N., & Hulme, D. (2012). *The role of NGOs and civil society in development and poverty reduction*. BWPI Working Paper 171. Brooks World Poverty Institute, University of Manchester.

Bonney, R., Shirk, J. L., Phillips, T. B., Wiggins, A., Ballar, H. L., Miller-Rushing, A. J., & Parrish, J. K. (2014). Next steps for citizen science. *SCIENCE, 343*, 1436–1437.

Burgess, H., DeBey, L. B., Froehlich, H., Schmidt, N., Theobald, E. J., Ettinger, A. K., … Parrish, J. K. (2017). The science of citizen science: Exploring barriers to use as a primary research tool. *Biology Conservation*, *2018*, 113–120.

Chandler, M., See, L., Copas, K., Bonde, A. M. Z., López, B. C., Danielsen, F., … Turak, E. (2017). Contribution of citizen science towards international biodiversity monitoring. *Biological Conservation*, *213*, 280–294.

Cigliano, J. A., Meyer, R., Ballard, H. L., Freitag, A., Phillips, T. B., & Wasser, A. (2015). Making marine and coastal citizen science matter. *Ocean and Coastal Management*, *115*, 77–87.

Clarke, L., & Agyeman, J. (2011). Shifting the balance in environmental governance: Ethnicity, environmental citizenship and discourses of responsibility. *Antipod*, *43*(5), 1773–1800.

Dean, A. J., Church, E. K., Loder, J., Fielding, K. S., & Wilson, K. A. (2018). How do marine and coastal citizen science experiences foster environmental engagement? *Journal of Environmental Management*, *213*, 409–416.

Dhillon, C. M. (2017). Using citizen science in environmental justice: Participation and decision-making in a Southern California waste facility siting conflict. *Local Environment*, *22*(12), 1479–1496.

Dickinson, J. L., Shirk, J., Bonter, D., Bonney, R., Crain, R. I., Martin, J., … Purcell, K. (2012). The current state of citizen science as a tool for ecological research and public engagement. *Frontiers in Ecology and the Environment*, *10*(6), 291–297.

Dimick, A. S. (2015). Supporting youth to develop environmental citizenship within/against a neoliberal context. *Environmental Education Research*, *21*(3), 390–402.

Dobson, A. (2007). Environmental citizenship: Towards sustainable development. *Sustainable Development*, *15*, 276–285.

Edwards, B., & McCarthy, J. D. (2004). Resources and social movement mobilisation. In D. A. Snow, S. A. Soule, & H. Kriesi (Eds.), *The Blackwell companion to social movements* (pp. 116–152). London: Blackwell.

Ellis, R., & Waterton, C. (2004). Environmental citizenship in the making: The participation of volunteer naturalists in UK biological recording and biodiversity policy. *Science and Public Policy*, *31*(2), 95–105.

Fan, M. F. (2008). Environmental citizenship and sustainable development: The case of waste facility siting in Taiwan. *Sustainable Development*, *16*, 381–389.

Gebbels, S., Evans, S. M., & Delany, J. E. (2011). Promoting environmental citizenship and corporate social responsibility through a school/ industry/ university partnership. *Journal of Biological Education*, *45*(1), 13–19.

Grace-McCaskey, C. A., Iatarola, B., Manda, A. K., & Etheridge, J. R. (2019). Eco-ethnography and citizen science: Lessons from within. *Society and Natural Resources*, *32*(10), 1123–1138.

Gregory, A. J., & Atkins, J. P. (2018). Community operational research and citizen science: The two icons in need of each other? *European Journal of Operational Research*, *268*, 1111–1124.

Islam, M. R., & Morgan, W. J. (2012). Agents of community empowerment? The possibilities and limitations of non-governmental organisations in Bangladesh. *Journal of Community Positive Practices*, *4*, 703–725.

Ismail, R., Shaw, B., & Ooi, G. L. (2009). *Southeast Asian culture and heritage in a globalising world: Diverging identities in a dynamic region*. England: Routledge, Taylor and Francis Group.

Iverson, E., & Jóndóttir, G. (2019). 'We did see the lapwing' – practising environmental citizenship in upper-secondary science education. *Environmental Education Research*, *25*(3), 411–421.

James, H. (2019). Introduction: The dynamics of sustainability and environmental governance in the Asia Pacific. In H. James (Ed.), *Population, development and the environment: Challenges to achieving the sustainable development goals in the Asia Pacific* (pp. 1–10). Canberra: Palgrave Macmillan.

Karlsson, R. (2012). Individual guilt or collective progressive action? Challenging the strategic potential of environmental citizenship theory. *Environmental Values*, *21*(4), 459–474.

Liu, Q., Wang, R. Y., & Dang, H. (2018). The hidden gaps in rural development: Examining peasant-NGO relations through a post-earthquake recovery project in Sichuan, China. *The China Quarterly*, *233*, 43–63.

Pandya, R. E. (2012). Citizen science – new pathways to public involvement in research. *Frontiers in Ecology and the Environment*, *10*(6), 314–317.

Parrish, J. K., Burgess, H., Weltzin, J. F., Fortson, L., Wiggins, A., & Simmons, B. (2018). Exposing the science in citizen science: Fitness to purpose and intentional design. *Integrative and Comparative Biology*, *58*(1), 150–160.

Rahman, S. (2016). *Orienting ASEAN towards its people: Enabling engagement with local NGOs.* RSIS Working Paper No. 298. S. Rajaratnam School of International Studies, Singapore.

Rahman, S. (2019). *Living heritage of Mukim Tanjung Kupang.*, Johor, Malaysia: Kelab Alami.

Renkema, A., & Kinlan, D. (2000). Tanjung Pelepas Port: From jungle to Malaysia's newest container port. *Terra et Aqua*, *80*, 3–11. Accessed 30 June 2019, from https://www2.iadc-dredging.com/wp-content/uploads/2017/03/article-tanjung-pelepas-port-from-jungle-to-malaysia-newest-container-port-80-01.pdf

Riesch, H., & Potter, C. (2014). Citizen science as seen by scientists: Methodological, epistemological and ethical dimensions. *Public Understanding of Science*, *23*, 107–120.

Schild, R. (2016). Environmental citizenship: What can political theory contribute to environmental education practice? *The Journal of Environmental Education*, *47*(1), 19–34.

Scott, L. (2006). *Chronic poverty and the environment: A vulnerability perspective. CPRC working paper 62*. London: Overseas Development Institute.

Smith, M. J., & Pangsapa, P. (2008). *Environment and citizenship: Integrating justice, responsibility and civic engagement*. London: Zed Books.

Theobald, E. J., Ettinger, A. K., Burgess, H. K., DeBey, L. B., Schmidt, N. R., Froehlich, H. E., … Parrish, J. K. (2015). Global change and local solutions: Tapping the unrealised potential of citizen science for biodiversity research. *Biological Conservation*, *181*, 236–244.

Trejos, B., Huang, W. C., & Chiang, N. (2008). The role of NGOs in the development of community-based rural tourism in Costa Rica. *Journal of Rural Tourism Research*, *2*(1), 63–76.

Unit Perancangan Ekonomi Johor (UPEN). (2016). *Johor sustainable policy 2017*-2021. Iskandar Puteri, Johor, Malaysia, p.70.

# Index